NANOTECHNOLOGY SCIENCE AND TECHNOLOGY

ADVANCED COMPUTATIONAL TECHNIQUES IN NANOSCIENCE

NANOTECHNOLOGY SCIENCE AND TECHNOLOGY

Additional books in this series can be found on Nova's website under the Series tab.

Additional e-books in this series can be found on Nova's website under the e-book tab.

NANOTECHNOLOGY SCIENCE AND TECHNOLOGY

ADVANCED COMPUTATIONAL TECHNIQUES IN NANOSCIENCE

A. K. HAGHI
F. NAGHIYEV
AND
S. ABDULLAYEVA

New York

For permission to use material from this book please contact us:
Telephone 631-231-7269; Fax 631-231-8175
Web Site: http://www.novapublishers.com

Library of Congress Cataloging-in-Publication Data

ISBN: 978-1-62257-791-0

Published by Nova Science Publishers, Inc. † New York

CONTENTS

About the Editors vii

Preface ix

Introduction xi

Chapter 1 **Properties of Carbon Nanotubes** 1

1.1. Structures of a Crystal Lattice of Diamond and Graphite 1

1.2. Graphene 3

1.3. Carbon Nanotubes 4

1.4. Fullerenes 6

1.5. Classification of Nanotubes 7

1.6. Chirality 8

1.7. Diameter, Chirality Angle and the Mass of Single-walled Nanotube 12

Chapter 2 **Micro and Nanoscale Systems** 17

2.1. Introduction 17

2.2. Electro-kinetic Processes in Micro and Nanoscale Systems 19

2.3. Continuum Hypothesis 20

2.4. The Molecular Dynamics Method 22

2.5. Van der Waals Equation. Corresponding States Law 26

Chapter 3 **Rheological Properties and Structure of the Liquid in Nanotubes** 33

3.1. Slippage of the Fluid Particles Near the Wall 33

3.2. The Density of the Liquid Layer Near a Wall of Carbon Nanotube 36

3.3. The Effective Viscosity of the Liquid in a Nanotube 38
3.4. The Release of Energy due to the Collapse of the Nanotube 43

Chapter 4 Fluid Flow in Nanotubes 47
4.1. Introduction 47
4.2. Modeling in Nanohydromecanics 57

Chapter 5 Nanohydromechanics 71
5.1. Nanophenomenon in Oil Production 71
5.2. Petroleum Composition 72
5.3. Nanobubbles of Gas, Oil in the Pore Channels and Water 78
5.4. Properties of Aluminum 88
5.5. The Use of Aluminum Powder in Oil Production 88

References 91

Index 97

About the Editors

A. K. Haghi, PhD, holds a BSc in urban and environmental engineering from University of North Carolina (USA), a MSc in mechanical engineering from North Carolina A&T State University (USA), a DEA in applied mechanics, acoustics and materials from Université de Technologie de Compiègne (France) and a PhD in engineering sciences from Université de Franche-Comté (France). He is the author and editor of 65 books, as well as 1000 published papers in various journals and conference proceedings. Dr. Haghi has received several grants consulted for a number of major corporations and is a frequent speaker to national and international audiences. Since 1983, he served as professor at several universities. He is currently Editor-in-Chief of the International Journal of Chemoinformatics and Chemical Engineering and Polymers Research Journal and on the editorial boards of many International journals. He is also faculty member of University of Guilan (Iran) and a member of the Canadian Research and Development Center of Sciences and Cultures (CRDCSC), Montreal, Quebec, Canada. Contact information: e-mail: Haghi@Guilan.ac.ir.

F. B. Nagiyev holds a M.Sc. in mechanics - mathematics from Moscow State University (Moscow, Russia), a PhD in mathematics and physics from Moscow State University (Moscow, Russia) and Doctor of Sciences in mathematics and physics from Leningrad Polytechnic Institute (Leningrad, Russia).

He is the author over 100 articles, 2 books and more than 200 analytical reports, projects on fundamental and applied studies in hydro-mechanics and thermal physics of multiphase media, mathematics, research studies in oil and gas sphere, power-engineering and technological complexes, education and sciences.

F. B. Nagiyev is member of Academician of International Information Academy (Moscow, Russia) and Academician of International Eco-Energy Academy (Baku, Azerbaijan).

F. B. Nagiyev is a Leading Research Officer of Ministry of Communications and Information Technologies of Azerbaijan Republic. Contact information: Baku, AZ1022, 96/A Mardanov Gardashlari Street, Apt.40, Phones: (+99412) 439-35-10 (office), (+99412) 597-27-59 (home), (+99450) 354-13-60 (cell), email: nagiyevfaik @gmail.com.

S. Abdullayeva holds a M.Sc. in physics from Azerbaijan State University, (Baku, Azerbaijan), a PhD in mathematics and physics from Azerbaijan State University (Baku, Azerbaijan) and Doctor of Sciences in mathematics and physics from Institute of Applied Physics, (Ministry of Defense Industry of the Former Soviet Union) Moscow (Russia).

She is the author over 130 articles, thirteen patents registered in Moscow.

S. Abdullayeva is member of American University Women Association, Indiana. 1994-1997, member of USA-Azerbaijani Friendship and Cultural Association. Washington, USA. 1995, member of Physical Society of The Former Soviet Union . 1983-1990, member of European Association of Higher Education- 2005-2006, member of the Supreme Attestation Commission attached to the President of the Republic of Azerbaijan – 2006- 2009. Contact information: Institute of Physics, National Academy of Sciences, Baku, Azerbaijan, 33, G.Javid Ave., AZ-1143, Baku, Azerbaijan, Phones: (+99412) 4394057 (office), (99450)2110686 (cell).

PREFACE

The results of investigation of liquid dynamics in carbon nanotube are presented. The analysis of a liquid structure and character of its flow plus the results of experiments show that the simulation of fluid flow for nanoscale systems should be based on the continuum hypothesis taking into account the quantized character of the liquid in the length scale of intermolecular distances.

Consideration of the flow characteristics allows to construct the analogy of behavior of the liquid in a nanotube with a flow of a viscoplastic Bingham fluid.

A model of mass transfer of liquid in a nanotube is introduced – it is based on the possibility of forming an empty interlayer between the moving fluid particles and the particles of the wall of the nanotube.

The book is prepared in the Research Center of High Technologies. It is approved to the edition at session of Scientific Council of the Center on September, 13th, 2011.

Reviewers: Vitaliy Solomonovich Ginzburg, Director of GEDA and TDN EOD Development Thomson Reuters Markets, New York, USA, Jack Petrovich Dvorkin, Senior Research Scientist - Rock Physics, Department of Geophysics, Stanford University, Stanford, CA, USA.

Approved by the Ministry of Communications and Information Technologies of the Azerbaijan Republic for use in educational process by students of higher educational institutions, post-graduate students, and also scientists.

Authors: A.K.Haghi, F.Naghiyev, S.Abdullayeva, 2012.

INTRODUCTION

Important developments were made in recent years in the new field of fluid mechanics associated with nanotechnology – Nanohydromechanics.

In general nanotechnology is a set of methods for production of items with a given atomic structure by manipulating atoms and molecules. Traditional hydrodynamics is associated with gases and liquids motion at macroscopic scale. Micro- and Nanohydromechanics is the area of mechanics where one studies the motion of gas and liquids in amounts conventionally related to nanotechnology (less than 100 nm = 0.1 microns).

Internal fluid dynamics describes the gas flow in micro- and nanochannels and tubes (including nanotubes). Smallest diameter pipes in the nature are carbon nanotubes (CNTs). Medications are often delivered to the patient body through the micro holes (gramicidin ion channel has diameter of a pore as 0.4 nm and length as 2.5 nm).

The size and surface properties of nanoparticles that enter the body through respiratory system affect the way they are moving throughout the body. The study first published by U.S. scientists may be useful for both development of sanitary standards on working with nanoparticles and new drugs development.

The extent and method of purifying lungs from nanoparticles that entered via respiratory system depends on size and surface characteristics of nano objects. Akira Tsuda (Harvard School of Public Health) decided to study how nanoparticles created by incomplete combustion processes or industrial emissions are absorbed by the body and/or removed from it. Teaming up with John Frangioni's group Tsuda has studied how these nanoparticles travel throughout the body.

The researchers studied the behavior of various fluorescent nanoparticles that differed from each other in size and composition. These nanoparticles were injected into the lungs of rats and researchers used real-time fluorescence spectroscopy to observe how the particles are absorbed by the body, migrate through the organs of rodents, and how many such particles are excreted from the body within an hour after they entered into the body.

It was found that most of the nanoparticles remain in the lungs, however if the diameter of nanoparticles is less than 34 nanometers – they can migrate to the lymph nodes. The speed of such movements depends on their size. Larger nanoparticles migrate more slowly. Speed of movement also depends on surface characteristics. Nanoparticles with zwitter-ionic, anionic and polar surface migrated to the lymph nodes, while nanoparticles with cationic surface remained in the pulmonary cells. Nanoparticles with zwitter-ionic surface with the size lesser than six nanometers moved quickly into the bloodstream and then got excreted by the kidneys.

Thus, it is clear that the fluid flow through micro- and nano-tubes represents a fundamental interest for many biological and technical devices and systems. Therefore flows in the nanometer size channels are intensively studied currently.

Importance of modeling in nanohydromechanics is supported by the results of numerous experiments conducted over the past two decades that revealed significant differences in the behavior of fluids in volumes with dimensions of about 10 of molecular diameters or less from the predictions of the classical continuum theories. The analysis shows that in microtubules of 50 nm in diameter the flow is continuous while in microtubules with the diameter of 5 nm it is not continuous, i.e. there is strong difference in the fluid – wall interaction in the range of 5-50 nm.

Increased carrying ability of the membranes of carbon nanotubes. Carbon nanotube (CNT) is a channel whose diameter is several times greater than the characteristic size of atomic particles, thus CNT can be considered as a reservoir for storage of gaseous and liquid substances. Recently the joint experiment of two groups – Lawrence Livermore Natural Laboratory and University of California, Berkeley (USA) – showed that nanotubes could serve as a channel for the transport of such substances with a capacity of 2-3 orders of magnitude higher than the corresponding quantities determined by the classical gas dynamics.

In the experiment, the film of closely packed ($\sim 2{,}5 \cdot 10^{11}$ $cм^{-2}$) of vertical double-layer carbon nanotubes were grown on a silicon chip by chemical vapor precipitation in the presence of a catalyst. The space between the tubes filled with silicon nitride (Si_3N_4) to gas or liquid is passed only through the inner cavity of nanotubes. Excess of silicon nitride was removed from both ends of the chip by ion milling, which resulted in a nanotube with both ends opened up. Measurements made during the passage through the nanotube of colloidal gold particles of various sizes, showed that these membranes are able to pass the particles with lateral dimensions between 1.3 and 2 nm.

Bandwidth ability of obtained membranes were determined for water as well as for the following gases - H_2, He, Ne, N_2, O_2, Ar, CO_2, Xe, CH_4, C_2H_6, C_3H_6, C_4H_6, C_4H_8. Measurements were made in the Knudsen regime. The ratio of the characteristic path length of gas molecules to the diameter of nanotubes is much greater than unity and ranges from 10 to 70. The bandwidth ability of the membrane for various gases was not the same.

Similar properties of membranes found scientists from the Lawrence National Laboratory and University of California (USA) under the leadership Olgica Bakajin, which allowed them to create a quick filter of nanotubes (Bakajin Olgica et al, 2001), (Chou C.-F. and et al 2003).

Scientists demonstrated the filter element where nanotubes play the role of membrane. It was planned to use for a study of transport of gases and fluids through the carbon nanotubes with diameter less than two nanometers. Based on previously developed mathematical model it was assumed that the rate of passage of substances through the nanotube has to be very high. It turned out to be true.

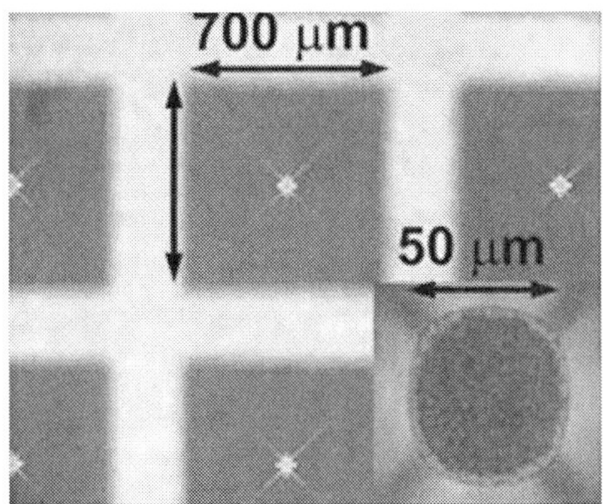

Figure 1.

Core of nanofilter is a matrix of vertically arranged double-layer nanotubes synthesized by precipitation in vapor phase on silicon nitride substrate (figure 1). Having free space filled by silicon nitride and ends of nanotubes opened up by etching - researchers obtained the filter element.

Due to the high density of nanopores per unit area of the nanotube, the physics of fluid dynamics is different from the classical one. As the scientists proved, the filter with pores density of $2{,}5 \cdot 10^{11}$ per cm^2 passes the fluid at a rate from 100 to 100,000 times faster than predicted by classical theory of liquids. This result allows the developing of commercially successful water and gases filtration and water desalination systems. Compared to polycarbonate filters, the nanotubes are characterized by lower limit of the passing particles.

The difference in the speed of passage of gas through the membrane of different masses of the CNT indicates the selectivity of the migration process. This can be used for solving of tasks of the gases separation of different sorts or different isotopic modifications. Multiple excess overlook of membranes based on carbon nanotubes over the value characteristic of the Knudsen regime, is due to the changing nature of the interaction of gas molecules with the inner walls of the nanotubes compared to the macroscopic surface.

The inner surface of the nanotubes is smooth on length scales down to atomic ones. At the same time the macroscopic surfaces of porous materials have a roughness on much larger scale. For this reason, the nature of the interaction of atomic particles with the walls of the nanotubes largely corresponds to mirror reflection and not diffuse reflection, as is in the case of macroscopic surfaces. Thus the gas inside CNT feels much less resistance from the surface than is determined by classical expression for the Knudsen flow.

Membranes based on carbon nanotubes are able to pass not only gases but also liquid substances. In this experiment showed that the overlook properties of membranes for water are more than three orders of magnitude higher than the corresponding value calculated on the basis of the classical formula of Hagen-Poiseuille. This effect is also related to the difference of the interaction of liquid with the inner walls of CNTs compared to the macroscopic surface. Liquid undergoes glide over the surface of CNTs, so in this case, no longer fulfilled traditionally used boundary conditions under which the flow velocity at the wall is zero.

Features of micro - and nanohydromechanics: a very large ratio of surface to volume measurements; comparable sizes of channel itself and molecules moving through the channel; possibility of significant density fluctuations in contrast to makroflows; transport properties (viscosity, diffusion, thermal conductivity) might contain the size factor (as in turbulence); nanoflow interaction with the wall might be the determining factor; there is no any exact boundary conditions; a continuum approximation can be violated; some events observed in micro- and nanoflows just do not exist in macro-hydrodynamics.

At the nanometer scale liquid shows unusual properties, for example, an abrupt increase in viscosity and density near the walls of nanocapillaries, the change of thermodynamic parameters of the liquid, as well as atypical chemical activity at the border between solid and liquid phases. The experiments discovered significant increase in the effective viscosity of the fluid as compared to its macroscopic value. This viscosity depends on the nanotube diameter.

The effective viscosity of the liquid in a nanotube is determined as follows. Let's compare two nanotubes of the same size and under the same pressure – one filled by liquid possibly containing crystallites and another filled by liquid considered as a homogeneous medium (i.e. without considering the crystallite structure) where Poiseuille flow is realized. The viscosity of a homogeneous fluid, that ensures the same mass transport for both tubes is the effective viscosity of the flow in the first nanotube.

Channel width reduction leads to the increase of Knudsen number and the raise of surface interactions role. And macroscopic consideration of natural gas as a continuous medium becomes invalid. Therefore it is accepted to use microscopic approach based on the methods of kinetic theory, direct statistical and molecular dynamic simulations.

Classical hydrodynamics, not taking into account atomic (molecular) structure of the liquid, does not adequately describe the flow of liquids in nanochannels with a width of about 10 molecular diameters or less. It is a major problem for nanotechnology researchers – the laws of classical physics no longer apply.

The next question is related to the use of classical or quantum hydrodynamics for describing nanoflow. Quantum hydrodynamics is important where the laws of classical physics become broken. And it is clear that due to much smaller scale the number of such problems in the micro- and nanohydromechanics is significantly bigger than in classical hydrodynamics, in particular, in a heat transfer from fluid to the wall (phonons) and phenomena that involves electrons (electromagnetic phenomena).

As a rule, the factors affecting nanotechnology applications are complex, i.e. electro-, hydraulic-, magnetic, optical and other processes are all important. To solve quantum hydrodynamics problems one would take into account the processes where quantum effects are important (photons and electrons) that have a minimum mass, because they correspond to the greatest lengths of de Broglie, that is important are the quantum phenomena in hydrodynamics.

The practical use of nanotubes. One of the most attractive destinations for using nanotubes is microelectronics. Small size, possibility to obtain the needed electrical conductivity during synthesis, mechanical strength and chemical stability make nanotubes very desirable material for microelectronics work elements.

Nanotubes have a number of interesting properties. For example, two nanotubes of the same chemical composition might have different conductivity depending on their configurations of atoms in the "lattice" (one to be closer to "metallic" and the other to "solid state").

Theoretical calculations have shown that if the defect in the form of a pentagon-heptagon pair is embedded into an ideal single-walled nanotube with chirality (8, 0) - the chirality of the tube in the vicinity of the defect become (7, 1). A nanotube with chirality (8, 0) is a semiconductor with a band gap of 1.2 eV, whereas the nanotube with chirality (7, 1) is a semimetal with zero band gap. Thus, a nanotube with an embedded defect can be considered as a hetero-conducting super small metal-semiconductor element.

Currently, the efforts of scientists are focused on developing technologies to obtain carbon nanotubes filled in with conducting or superconducting material. The goal is to create conductive compounds – the base for production of the smallest nanoelectronic devices.

Individual nanotubes can be used as delicate probes for the study of surfaces with roughness on the nanometer level. In this case the extremely high mechanical strength of the nanotube is utilized. The modulus of elasticity E along the longitudinal axis of the nanotube is about 7000 GPa, whereas the probes made of steel and iridium barely reach the values of E=200 and 520 GPa, respectively. In addition single-walled nanotubes, for example, may get elastically lengthened by 16%. To visualize such a material property from 30 cm iron spoke, it should be lengthened by 4.5 cm under load and then get back to the original length. The probe of the nanotube with superelastic properties will bend elastically with the load exceeding certain level, thereby providing contact with the surface.

High modulus of elasticity of carbon nanotubes allows creation of composite materials having high strength during ultra high elastic deformation.

And it can be successfully used to produce ultra-light and heavy-duty fabric for firefighters and astronauts clothing.

The strength of a material is its ability to withstand an applied stress without failure – either fracture (separation of the parts) or irreversible change in shape (plastic deformation). When a cylindrical sample with a cross section of S get stretched by force F - it deforms elastically at first (reversible deformation, see the section ***O*** on the curve in figure 2), and then plastically, i.e. irreversible (see section ***П*** figure 2).

At the top of figure 2 you can see a schematic drawing of a red cylindrical sample with a cross-sectional area S under applied force F getting extended by an amount of $L - L_0$ where L_0 is the original sample length. At the bottom there is a drawing of relationship between mechanical pressure and relative deformation with arrows indicating the results for titanium, steel and bronze.

While sample gets deformed its structural inhomogeneities (lattice defects or dislocations) start moving, colliding with others, thus forming micro cracks. The more dislocations exist in the sample and the sooner they start moving, the more micro cracks get formed.

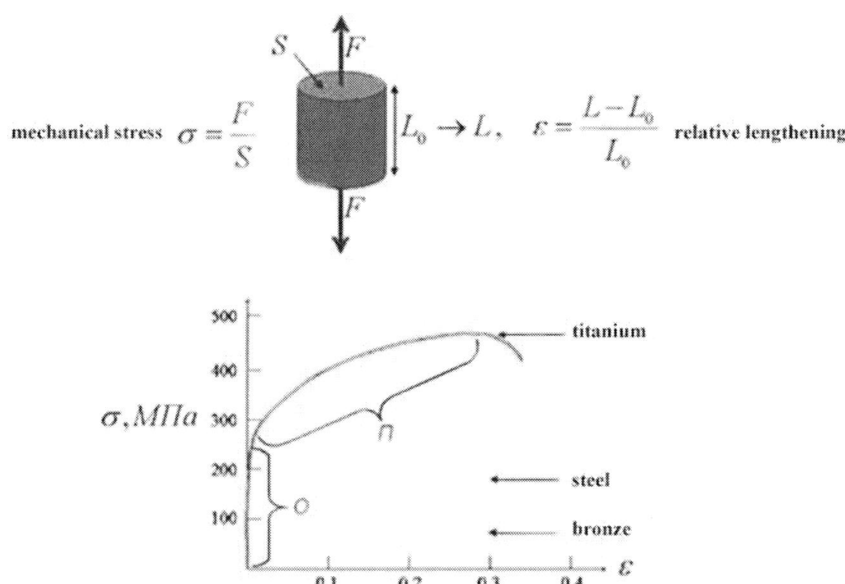

Figure 2. Schematic illustration of red cylindrical sample and the relationship between mechanical stress and relative lengthening during tension specimen.

When σ ($\sigma = F/S$) reaches certain level then the neighboring micro cracks connect to each other reaching critical size and the sample gets destroyed.

Nanowire is a single crystal virtually free from defects (dislocations). In addition the surface of nanowire has very small radius of curvature (10 nm), it is highly compressed and therefore prevents the movement of dislocations out, i.e. formation of micro cracks. All this leads to the fact that nanowire has almost no plastic deformation and its tensile strength is ten times higher than for normal samples (see figure 3).

Figure 3 shows the relationships between stress and relative deformation during tensile tests of various diameters micro samples made from Ni and its alloys $Ni_3Al - Ta$ (Uchic et al, 2004). The diameter values are shown next to corresponding curves. For comparison, the red shows the "stress – relative strain" for macro samples. It shows that the tensile strength increases with decreasing diameter of wire.

Let's calculate the strength of carbon nanotube using single-wall of "zig-zag" nanotube (figure 4). We lock the invisible end of the tube and apply tensile force F to the other end. Figure 4 shows blue lines crossing C - C links oriented along the axis of the tube, and the yellow arrow shows the direction of the tensile force F.

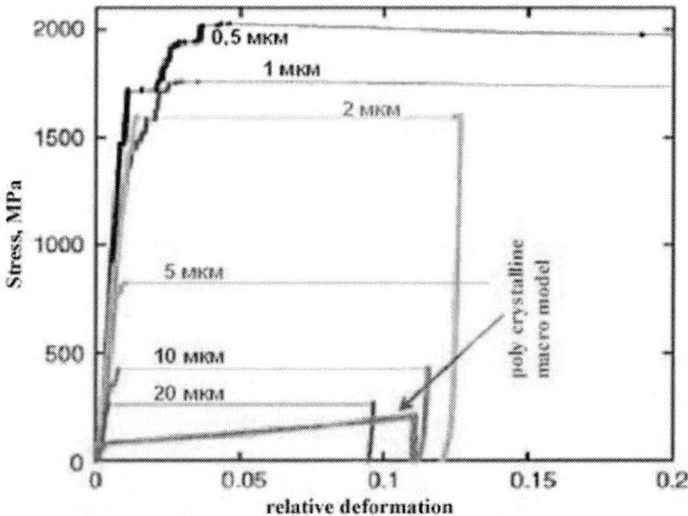

Figure 3. The relationship between stress and relative deformation during tensile microsamples of different diameters of Ni and its alloys $Ni_3Al - Ta$.

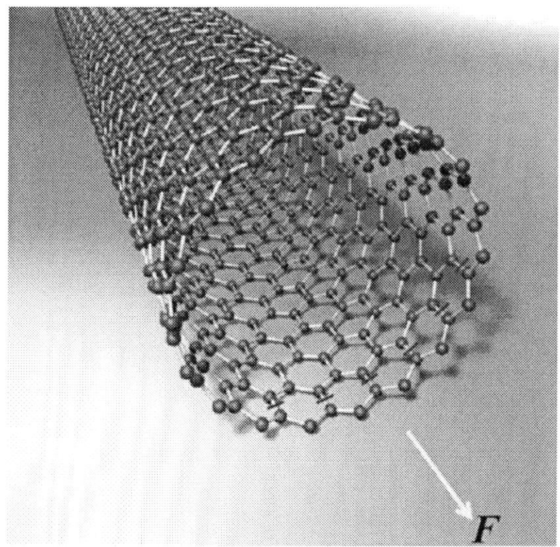

Figure 4. Carbon nanotube schema.

Let carbon atoms form the identical connections (C-C, σ -connections) in the nanotube and the angles between them equal 120^0. Then, when nanotube gets stretched – these connections will stretch the same way. And nanotube can explode in a most unexpected way depending, for instance, which C-C connection breaks first.

To simplify the calculations let's assume that the tension breaks only C-C connections oriented along the axis of the tube and located on the same cross-section (blue break lines on figure 4).

It is known that the distance between adjacent carbon atoms in the nanotube is approximately equal to d =0.15 nm. It is easy to show that if the tube diameter is D, then the amount of N connections oriented along the axis of the tube is:

$$N = \frac{\pi D}{\sqrt{3}d} \qquad (1)$$

Value of force applied to each C-C connections equals F / N.

The strength of C-C connection can be found from a curve on figure 5 that shows dependence of connection potential energy on the distance between atoms.

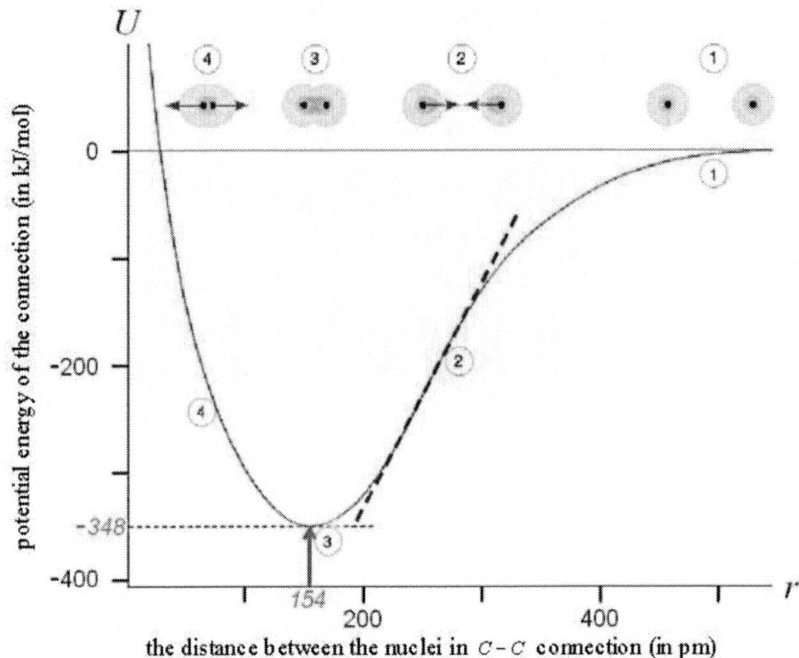

Figure 5. How potential energy of $C-C$ connection mole (i.e. 6.1023) depends on the distance between the nuclei.

Figure 5 diagram shows that the potential energy of connection reaches the minimum when the distance between the nuclei of atoms is 154 pm. This determines the distance between carbon atoms when nanotube is not stretched.

Tangent of slope of the right branch of the curve in figure 5 is proportional to the force F_1 needed to hold the atoms at a given distance r :

$$F_1 = \frac{\partial U}{\partial r} \cdot \frac{1}{N_A},$$

where N_A – Avogadro's number, $6 \cdot 10^{23} mole^{-1}$.

To increase the distance between the carbon atoms one has to apply force F_1, and if it is greater than the maximum tangent of the slope (see blue dotted line in figure 5), then C-C connection breaks. It happens when

$$F_1 > \frac{348 \cdot 10^3}{154 \cdot 10^{-12} \cdot 6 \cdot 10^{23}} = 3{,}8 \text{ нН} \qquad (2)$$

The nanotube will break when the force F stretching the tube becomes more than 3.8 N nN, where N is the number of C-C connections parallel to the axis of a single cross-section of the tube. If nanotube diameter D = 1.5 nm then N = 18 (from (1)). Therefore, the nanotube will break at F_{max}> 69 nN.

In order to calculate the strength σ_{max} of the nanotube, let's divide F_{max} by the cross-sectional area number $S = \pi D^2 / 4$:

$$\sigma_{max} = \frac{F_{max}}{\pi D^2} \cdot 4 = \frac{4 \cdot 69 \cdot 10^{-9}}{3{,}14 \cdot 2{,}25 \cdot 10^{-18}} = 39 \text{ ГПа} \qquad (3)$$

This value is quite close to the experimentally obtained values maximum (63 GPa) and, as expected, much more than the strength of the most strong types of steel (0.8 GPa).

Note that multiwalled nanotube strength is several times higher!

Another attractive nanotube property is its very high surface area. In the process of growth randomly oriented helical nanotubes get formed, that leads to a significant number of cavities and voids of nanometer size. As a result, the area per unit mass of nanotubes material reaches values of about $600\, m^2 / g$. Such a high surface area opens up the possibility of usage in filters and other devices of chemical technology.

The unique chemical properties of carbon nanotubes allow very small molecules of water to seep through them, whereas viruses, bacteria, toxic metal ions and toxic organic molecules cannot do this.

Thanks to the amazing properties, nanotubes can be used in various technological fields, ranging from the creation of composite materials and heavy-duty thread, to the construction of nanodevices and nanosensors.

Nanotubes can serve as the storage of hydrogen, the cleanest fuel.

The reserves of coal, oil and gas on Earth are limited. In addition, the burning of conventional fuels leads to the accumulation of carbon dioxide and other harmful pollutants in the atmosphere, and this in turn – to global warming, the signs of which humanity has already experienced. So today mankind is facing a very important problem - to replace traditional fuels in not so distant future.

It is really beneficial to use hydrogen - the most common chemical element in the universe – as a fuel. The oxidation (combustion) of hydrogen proceeds with the release of a very large amount of heat (120 kJ/kg) in addition to water. For comparison, the combustion of gasoline or natural gas releases three times less heat than that of hydrogen. It should also be noted that the combustion of hydrogen does not form any harmful to ecology output.

There are quite a few fairly cheap and environmentally friendly ways to produce hydrogen, however, hydrogen storage and transportation has been so far one of the unsolved problems. The reason for this is a very small size of hydrogen molecule. Hence hydrogen can leak through microscopic cracks and pores of conventional materials, and it might cause explosion. So the walls of containers for storing hydrogen should be thicker and therefore heavier. For more safety, it is better to freeze hydrogen to several dozen ^{o}K, which raises the price of storing even further.

The solution to the described above problem could become a device that plays the role of "sponge" - with ability to absorb hydrogen and hold it indefinitely. Obviously, such a hydrogen "sponge" must have a large surface area and chemical affinity to hydrogen. All these properties are present in carbon nanotubes.

It is well known all the atoms of carbon nanotube are on its surface. It is the base for one of the main mechanisms of how nanotube absorbs hydrogen – chemisorption, i.e. the adsorption of hydrogen H_2 on the surface of the tube with subsequent dissociation and formation of chemical connections of C-H. Absorbed in such a way, hydrogen can be extracted from nanotubes, for example, by heating it to $600\,^{o}C$. In addition, the hydrogen molecules are bound to the surface of the nanotubes due to physical adsorption by van der Waals interactions.

Researchers figured out that any efficient fuel cell based on hydrogen "sponge" should contain at least 63 kg of hydrogen per cubic meter. In other words, the weight of stored hydrogen must be not less than 6.5% of the weight of the "sponge." Currently the experimental “sponge” can store more than 18% of hydrogen, and it opens up broad prospects for development of hydrogen energy.

Suetin M.V. and Vakhrushev A.V. (2009) investigated nanocapsule for storing methane using methods of molecular dynamics. Nanocapsule consisted of three nanotubes (20,10) (10,10) and (8.8) joined together.

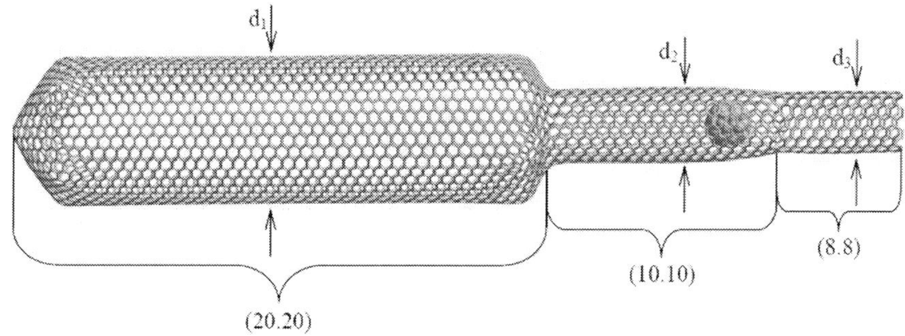

Figure 6. Nanocapsule for natural gas storage, consisting of three types of nanotubes (20,20), (10,10), (8,8) and containing $K@C_{60}$.

There was the endohedral complex $K@C_{60}$ with single positive charge inside nanocapsule that was used for blocking. Its movement was controlled by external electric field.

The modeling was done by molecular dynamics methods with 1fs step. The object of the study was nanocapsule whose structure is shown in figure 6.

The nanocapsule consists of "armchair" type nanotubes of different diameters: (20,20) - d_1=25.93 A=2.593 nano (1 angstrom=$10^{-10}\,м = 0{,}1н$), (10,10)-d_2=13,4 A and (8,8)-d_3=10,74 A, combined together by heptagonal rings ($d = 0{,}1295 \cdot (n,n)$). Nanocapsule contains endohedral complex $K@C_{60}$ where potassium atom has a single positive charge. The @ symbol means that K is contained inside C_{60}. The diameter d_2 of nanocapsule at the site (10,10) is big enough to let $K@C_{60}$ get in, however not that big to allow gas molecules to move between $K@C_{60}$ and the wall of nanocapsule. The diameter d_3 of nanocapsule at the site (8,8) is not sufficient for $K@C_{60}$ to pass, but good enough for methane molecules. Region of nanocapsules (20,20) is used to accumulate and store gas molecules. Endohedral complex $K@C_{60}$ in nanocapsule is moved by external electric field.

Nanocapsule works in 3 modes: adsorption, desorption and storage. At the stage of the adsorption $K@C_{60}$ is located near the base of nanocapsule (left end of the nanocapsule - figure 6). Methane molecules pass through the hole in the nanocapsule and get adsorbed on the walls. In order to switch to the storage mode it is necessary to seal the nanocapsule entrance. $K@C_{60}$ gets moved by electric field to block the entrance - now methane molecules can not leave the nanocapsule. It stops at (8,8) site unable to move any longer and stays there even after disappearance of the electric field - due to the pressure of methane molecules and capillary forces. At this stage (storage), the external thermodynamic conditions get back to normal.

To initiate the stage of desorption $K@C_{60}$ gets moved to the bottom of nanocapsule. Due to excess pressure inside the nanocapsule methane molecule starts leaving its inner space. However, significant part of methane remains in the nanocapsule concentrated in the areas of nanotubes (10.10) and (8.8). For complete extraction of methane $K@C_{60}$ gets moved to the site (10,10) and squeezes the methane out. Then $K@C_{60}$ gets back to the bottom of nanocapsule. Methane molecules are concentrated in the area of (10,10) again, and then $K@C_{60}$ squeezes them out. This happens as long as all of the methane molecule leave the interior of the nanocapsule.

The nanocapsule shown in Figure 6 contains 737 molecules of methane. The pressure inside the nanocapsule is about 10 MPa. The density of methane inside nanocapsules is ~ $82\,kg/m^3$. The weight content is calculated as follows:

$$W_t = \frac{N_{CH_4} \times m_{CH_4}}{N_{CH_4} \times m_{CH_4} + N_C \times m_C} \times 100\%,$$

where N_{CH_4} - number of methane molecules, N_C - number of carbon atoms in the nanotube, m_{CH_4} - mass of one molecule of methane and m_C - mass of one carbon atom. Thus, there is ~ 17.5 wt. % of methane in nanocapsule.

Figure 7 shows two graphs for desorption stage: the first is $K@C_{60}$ velocity V, and the second – the $K@C_{60}$ coordinate d. $K@C_{60}$ moves under an electric field influence from the site of nanotube (10,10) to the center of nanocapsule base – area of nanotube (20,20).

Strength of the electric field ($5{,}14 \cdot 10^9$ V/m) is determined by the need to overcome the capillary forces that oppose $K@C_{60}$ move from the site (10,10) to (20,20) plus the pressure of compressed methane.

Nanocapsule is similar to bottle-like pore so significant amount of methane molecules in the pore remains there even after the opening of the nanocapsule. In our case it is about 7.9 wt. % of methane remains in the nanocapsule. Given that the adsorption potential at the sites (10,10) and (8,8) is significantly higher than at (20,20), the methane molecules are concentrated in the first two sections. Now it is necessary to desorb the gas molecules. In order to do this, $K@C_{60}$ gets moved by electric field (magnitude $5{,}14 \cdot 10^9$ v/m is needed) toward the (10,10) site of the nanotube and squeezes methane from the nanocapsule.

In order to do this, $K@C_{60}$ gets moved by electric field (magnitude $5{,}14 \cdot 10^9$ v/m is needed) toward the (10,10) site of the nanotube and squeezes methane from the nanocapsule.

Figure 8 shows the curve of methane molecules desorption from nanocapsule. As you can see, it is almost linear while $K@C_{60}$ is moving from the site (10,10) and the desorption ends abruptly with $K@C_{60}$ reaching the site (8,8).

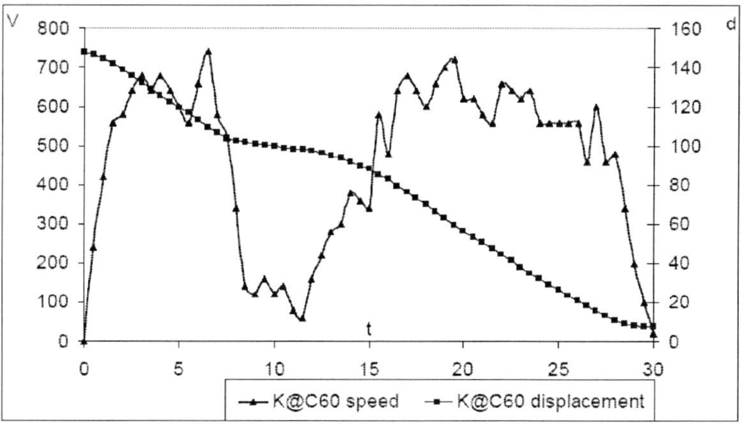

Figure 7. How $K@C_{60}$ velocity V and $K@C_{60}$ coordinate d behave during nanocapsule desorption.

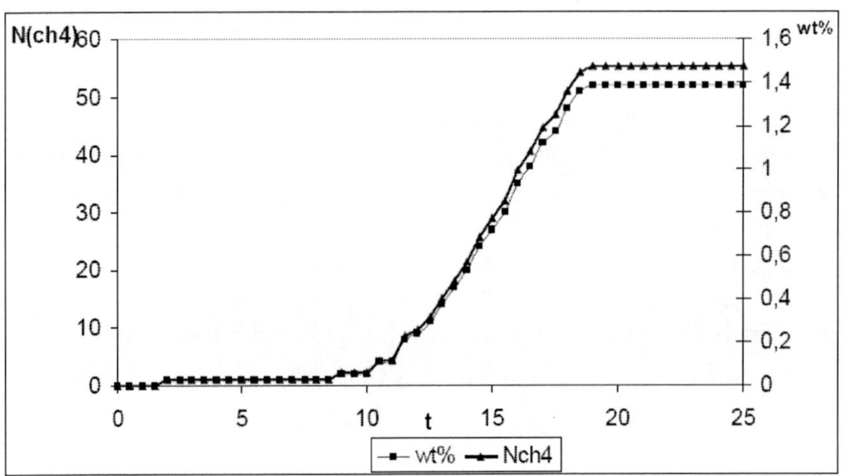

Figure 8. Quantity (N (ch4)) and weight (wt%) of methane molecules in the nanocapsule during $K@C_{60}$ extrusion.

In order to get $K@C_{60}$ back - the direction of the electric field should be reversed (with the same magnitude). So a single cycle of desorption removes about 1.4 wt. % of methane from the nanocapsule.

Meaning that if nanocapsule contains 7.9 wt. % of methane then the complete desorption is achieved by six cycles of $K@C_{60}$ extrusion. Of course, we should keep in mind that certain amount of methane might still remain in the nanocapsule, but higher adsorption potential of (10,10) area leads us to believe that the vast majority of methane molecules is concentrated precisely at (10,10) area.

As you can see the nanocapsule is fairly complex structure where the locking element $K@C_{60}$ behavior is controlled by an external electric field. And its direction and magnitude of tension determine the position of $K@C_{60}$ in nanocapsule and accordingly the phases of adsorption, desorption and storage.

Here are some possible applications of nanohydromechanics.

Microbubble medium (gas in liquid). This is an interesting area with multiple applications. And any success in the area requires significant efforts in several departments: studying of thermodynamics of two-phase systems with free boundary, as applied to micro- and nano-bubble environments; estimates of energy costs needed to obtain micro-bubble environments (there are various ways to do it); the kinetics of growth and destruction of micro-bubbles in the liquid; limits of tension leading to the destruction of micro-bubbles; physics of simple liquids, as applied to micro-bubbles mediums; development of mathematical models to describe the physical properties of micron- and nano-sized bubbles of media; theoretical methods and numerical modeling of viscosity, density and sedimentation stability of micro- and nano-bubble media; analysis of opportunities for modifying the properties of the liquid in hydrodynamic devices; carrying out theoretical studies of possible ways to get micro-bubbles nano-scale environments; atomic-force microscopy in nano-bubbles environments; nano-bubbles in sonoluminescence.

Micro-hydromechanics of oil. Behavior of oil in reservoirs at a depth of 1-3 km in hard rocks is determined by Darcy law. Typical values of permeability range from 5 to 500 mD. The permeability of coarse-grained sandstone is $10^{-8} - 10^{-9}\,sm^2$, the permeability of tight sandstone is around $10^{-10}\,sm^2$. For reservoirs with permeability as a fraction of micron the flow of oil is described by nanotechnology methods.

Microstructure of a viscoplastic fluid. In many cases, the liquid changes its properties due to irreversible processes occurring at micro- and/or nanoscale, thus there is a change in the rheological properties of liquids. Therefore, both experimental and theoretical studies of nanorheology gels used in hydrofracking are becoming more relevant.

Micro- and nanohydromechanics is a new area of fundamental and applied mechanics where fundamental research is the basis for creation of real world hydrodynamic nanotechnologies – the core of several promising XXI century technologies.

The main method of nanotubes manufacturing is arc discharge process. The anode "evaporates" and multi-layered nanotubes “grow” on the surface of the cathode.

Until now, nanotubes obtained for a variety of studies have not been able to provide the required purity of the experiment. The tubes have always contained some dirt making precise measurements impossible: the electron can be influenced by an additional electric potential caused by pollution making the results meaningless.

Chapter 1

PROPERTIES OF CARBON NANOTUBES

1.1. STRUCTURES OF A CRYSTAL LATTICE OF DIAMOND AND GRAPHITE

In 1991, Idzhima studied the sediments formed at the cathode during the spray of graphite in an electric arc. His attention attracted by the unusual structure of the sediment consisting of microscopic fibers and filaments. Measurements made with an electron microscope showed that the diameter of these filaments does not exceed a few nanometers and a length of one to several microns.

Having managed to cut a thin tube along the longitudinal axis, the researchers found that it consists of one or more layers, each representing a hexagonal grid of graphite, which is based on hexagon with vertices located at the corners of the carbon atoms. In all cases, the distance between the layers is equal to 0.34 nm, which is the same as that between the layers in crystalline graphite.

Typically, the upper ends of tubes are closed by multilayer hemispherical caps; each layer is composed of hexagons and pentagons, reminiscent of the structure of half a fullerene molecule.

The extended structure consisting of rolled hexagonal grids with carbon atoms at the nodes are called nanotubes.

Lattice structure of diamond and graphite are shown in figure 1.1. Graphite crystals are built of planes parallel to each other, in which carbon atoms are arranged at the corners of regular hexagons. The distance between adjacent carbon atoms (each side of the hexagon) $d_0 = 0{,}141\,nm$, between adjacent planes - 0.335 nm.

Each intermediate plane is shifted somewhat toward the neighboring planes, as shown in this figure.

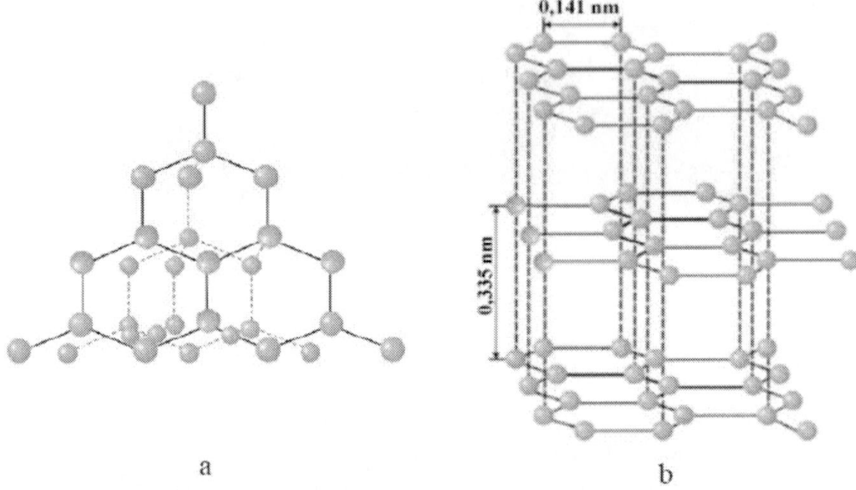

Figure 1.1. Structure of (a) diamond lattice and (b) graphite.

Figure 1.2. Schematic illustration of the graphene.

The elementary cell of the diamond crystal represents a tetrahedron, with carbon atoms in its center and four vertices. Atoms located at the vertices of a tetrahedron form a center of the new tetrahedron, and thus, are also surrounded by four atoms each, etc. All the carbon atoms in the crystal lattice are located at equal distance (0.154 nm) from each other.

Nanotubes are rolled into a cylinder (hollow tube) graphite plane, which is lined with regular hexagons (with carbon atoms at the vertices of a diameter of several nanometers) (Refer to Figure 1.2). Nanotubes can consist of one layer of atoms - single-wall nanotubes SWNT and represent a number of "nested" one into another layer pipes - multi-walled nanotubes - MWNT.

Nanostructures can be built from the molecular blocks as well. Such blocks or elements are graphene, carbon nanotubes and fullerenes.

1.2. Graphene

Graphene is a single flat sheet, consisting of carbon an atom linked together and forming a grid (each cell is like a bee"s honeycombs) (Refer to Figurc 1.2). Thc distancc bctwccn adjaccnt carbon atoms in graphcnc is about 0.14 nm.

Graphite, from which slates of usual pencils are made, is a pile of graphene sheets (figure 1.3). Graphenes in graphite is very poorly connected and can slide relative to each other. So, if you conduct the graphite on paper, then after separating graphene from sheet the graphite remains on paper. This explains why graphite can write.

Figure 1.3. Schematic illustration of the three sheets of grapheme.

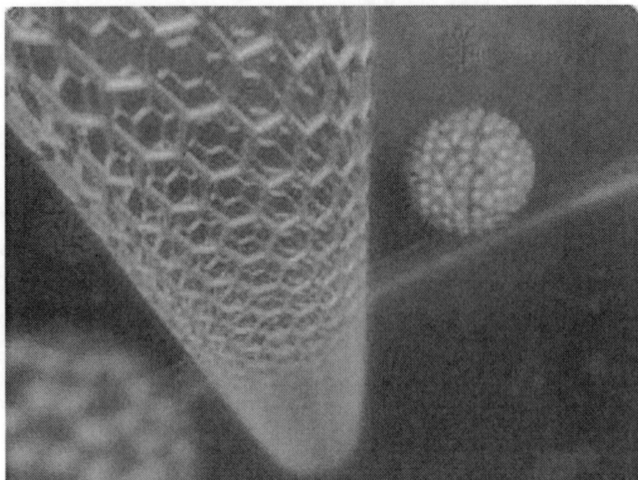

The way of folding nanotubes - the angle between the directions of nanotube axis relative to the axis of symmetry of graphene (the folding angle) - largely determines its properties.

Figure 1.4. Carbon nanotubes.

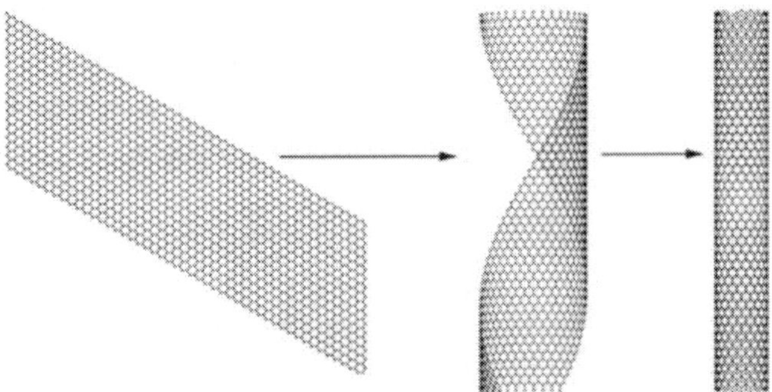

Figure 1.5. Formation of nanotube.

1.3. CARBON NANOTUBES

Many perspective directions in nanotechnology are associated with carbon nanotubes.

Carbon nanotubes are a carcass structure or a giant molecule consisting only of carbon atoms.

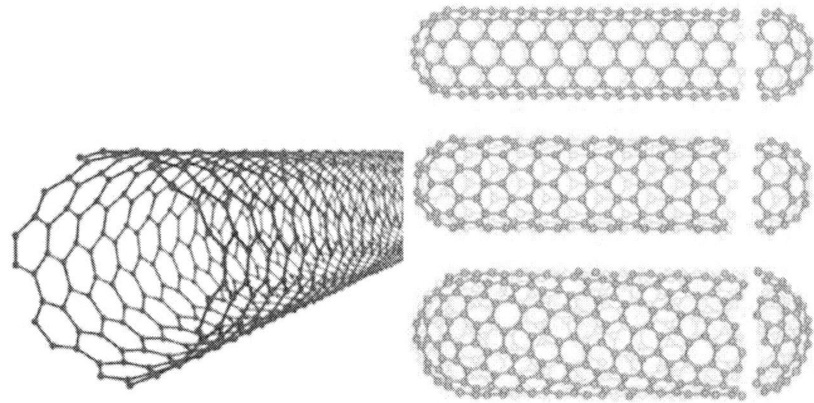

Figure 1.6. Left - schematic representation carbon nanotubes.

Carbon nanotube is easy to imagine, if we imagine that we fold up one of the molecular layers of graphite - graphene (figure 1.5).

The angle between the direction of nanotube axis relative to the axis of symmetry of graphene (the folding angle) determines its properties.

Nanotubes formed themselves, on the surface of carbon electrodes during arc discharge between them. At discharge, the carbon atoms evaporate from the surface, and connect with each other to form nanotubes (figure 1.6).

The diameter of nanotubes is usually about 1 nm and their length is a thousand times more, amounting to about 40 microns. They grow on the cathode in perpendicular direction to surface of the butt. The so-called self-assembly of carbon nanotubes from carbon atoms occures. Depending on the angle of folding of the nanotube they can have conductivity as high as that of metals, and they can have properties of semiconductors.

Carbon nanotubes are stronger than graphite, although made of the same carbon atoms, because carbon atoms in graphite are located in the sheets. And everyone knows that sheet of paper folded into a tube is much more difficult to bend and break than a regular sheet. That's why carbon nanotubes are strong. Nanotubes can be used as a very strong microscopic rods and filaments, as Young's modulus of single-walled nanotube reaches values of the order of 1-5 TPa, which is much more than steel! Therefore, the thread made of nanotubes, the thickness of a human hair is capable to hold down hundreds of kilos of cargo.

It is true that at present the maximum length of nanotubes is usually about a hundred microns (which is certainly too small for everyday use).

However, the length of the nanotubes obtained in the laboratory is gradually increasing.

1.4. FULLERENES

The carbon atoms, evaporated from a heated graphite surface, connecting with each other, can form not only nanotube, but also other molecules, which are closed convex polyhedra. In these molecules, the carbon atoms are located at the vertices of regular hexagons and pentagons which make up the surface of a sphere or ellipsoid.

All of these molecular compounds of carbon atoms are called fullerenes on behalf of the American engineer, designer and architect R. Buckminster Fuller, who used pentagons and hexagons for construction of domes of his buildings, (figure 1.7).

The molecules of the symmetrical and the most studied fullerene consisting of 60 carbon atoms (C_{60}) (Refer to Figure 1.8). The diameter of the fullerene C_{60} is about 1 nm.

Figure 1.7. Biosphere of Fuller (Montreal, Canada).

The image of the fullerene C_{60} many consider as a symbol of nanotechnology.

Figure 1.8. Schematic representation of the fullerene C_{60} .

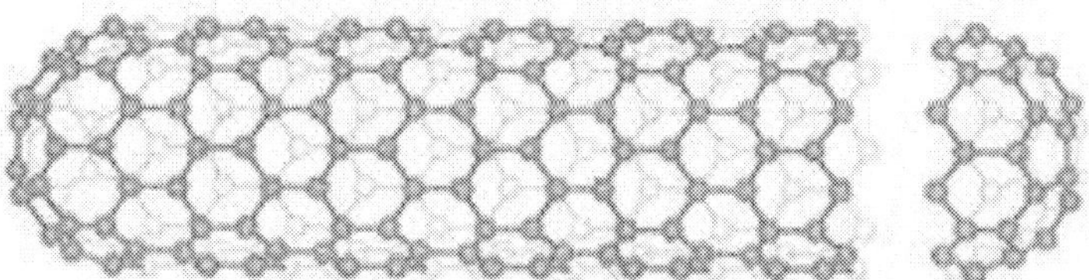

Figure 1.9. Graphical representation of single-walled nanotube.

1.5. CLASSIFICATION OF NANOTUBES

The main classification of nanotubes is conducted by the number of constituent layers.

Single-walled nanotubes –The structure of the nanotubes can be represented as a "wrap" hexagonal network of graphite (graphene). This is based on hexagon with vertices located at the corners of the carbon atoms in a seamless cylinder. The distance d between adjacent carbon atoms in the nanotube is approximately equal to $d = 0{,}15$ nm.

Multi-walled nanotubes consist of several layers of graphene stacked in the shape of the tube. The distance between the layers is equal to 0.34 nm, that is the same as that between the layers in crystalline graphite.

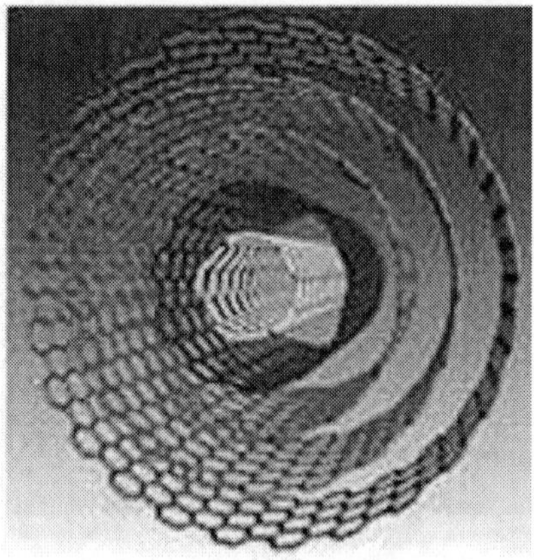

Due to its unique properties (high fastness (63 GPa), superconductivity, capillary, optical, magnetic properties, etc.), carbon nanotubes could find applications in numerous areas:

- additives in polymers;
- catalysts;
- absorption and screening of electromagnetic waves;
- transformation of energy;
- anodes in lithium batteries;
- keeping of hydrogen;
- composites (filler or coating);
- nanosondes;
- sensors;
- strengthening of composites;
- supercapacitors.

Figure 1.10. Graphic representation of a multiwalled nanotube.

1.6. CHIRALITY

Chirality; is a set of two integer positive indices (n, m), which determines how the graphite plane folds.

Values of parameters (n, m) are distinguished as;

- direct (achiral) high-symmetry carbon nanotubes

- armchair $n = m$
- zigzag $m = 0$ or $n = 0$

- helical (chiral) nanotube

Figure 1.11 show schematic representation of the atomic structure of graphite plane.

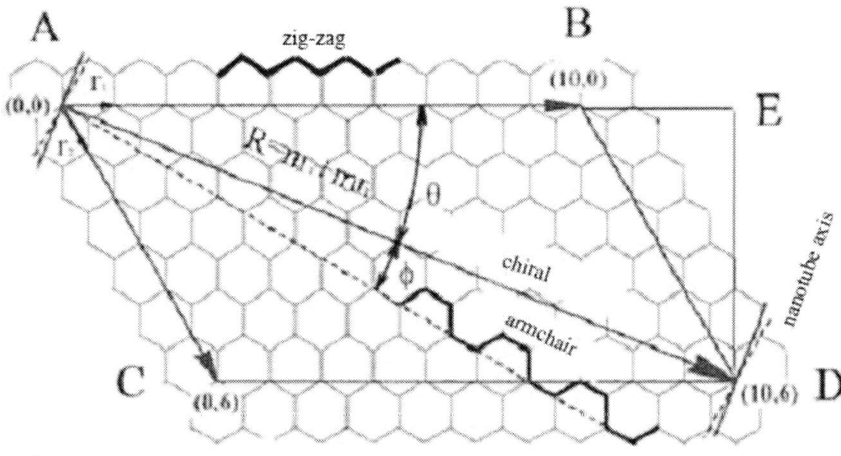

a.

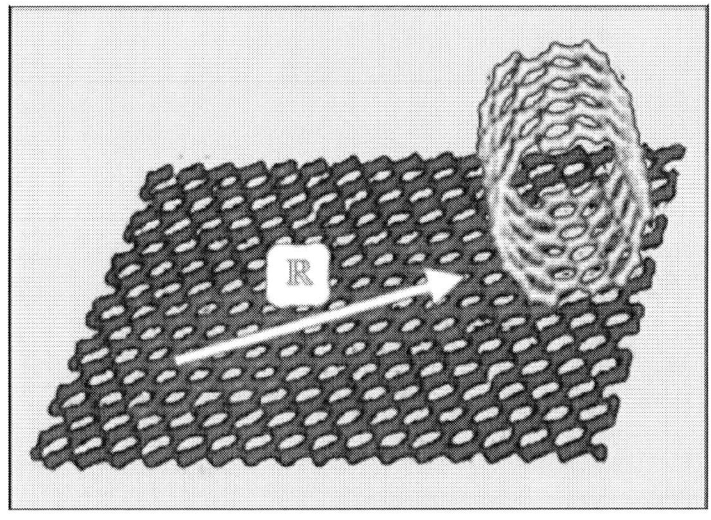

b.

Figure 1.11. Schematic representation of the atomic structure of graphite plane.

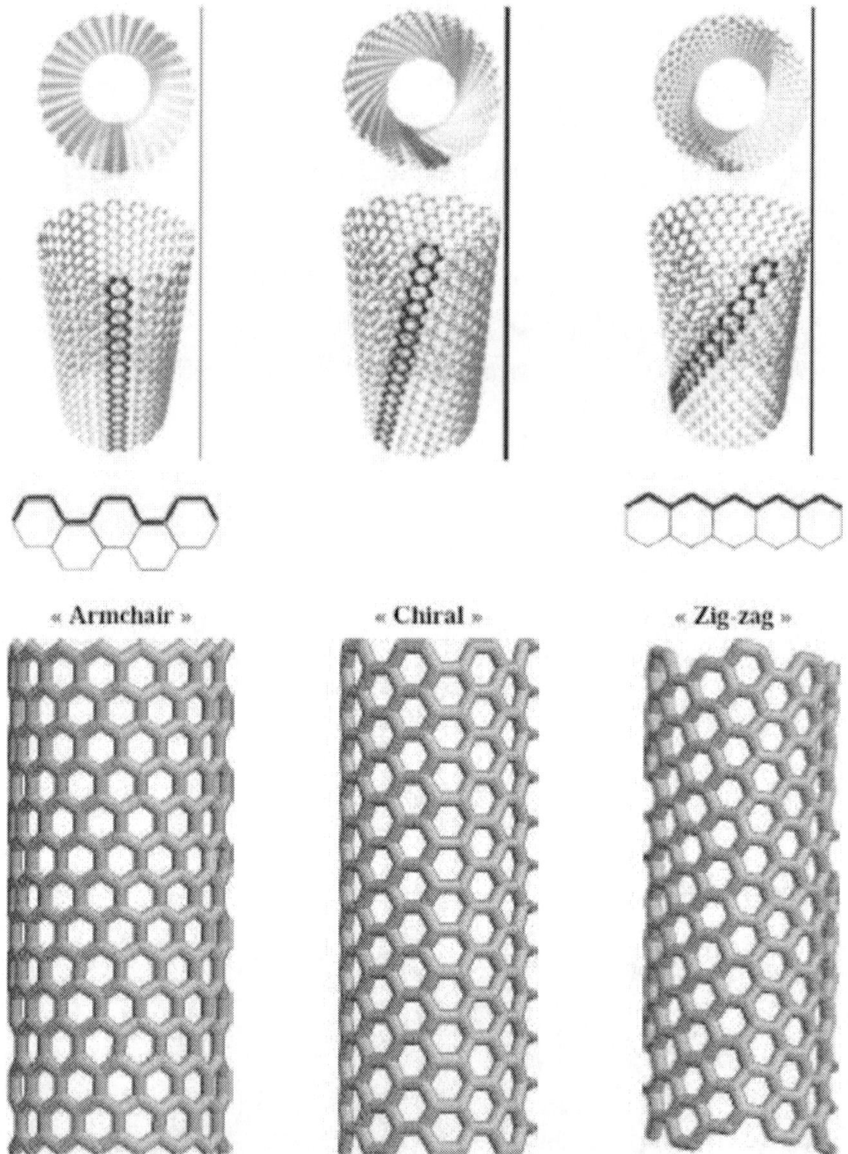

Left to right: the zigzag (16,0), armchair (8,8) and chiral (10,6) carbon nanotubes.

Figure 1.12. Single-walled carbon nanotubes of different chirality (in the direction of convolution).

The cylinder is obtained by folding this sheet. To obtain a carbon nanotube from a graphene sheet, it should turn so that the lattice vector $\overline{R}$ has a circumference of the nanotube (Figure 1.11b). This vector can be expressed in terms of the basis vectors of the elementary cell graphene sheet $\vec{R} = n\vec{r_1} + m\vec{r_2}$. Vector $\overline{R}$, which is often referred to simply by a pair of indices (n, m), called the chiral vector. It is assumed that $n > m$. Each pair of numbers (n, m) represents the possible structure of the nanotube.

In other words the chirality of the nanotubes (n, m) indicates the coordinates of the hexagon, which as a result of folding the plane has to be coincide with a hexagon, located at the the beginning of coordinates (figure 1.12).

Many of the properties of nanotubes depend on the value of the chiral vector. For example, a nanotube (10,10) in the elementary cell contains 40 atoms and is the type of metal. Nanotube (10, 9) in 1084 is a semiconductor (figure 1.13).

If the difference $n - m$ is divisible by 3, then these CNTs have metallic properties. Semimetals are all achiral tubes such as "chair". In other cases, the CNTs show semiconducting properties. The chair CNTs $(n = m)$ are strictly metal.

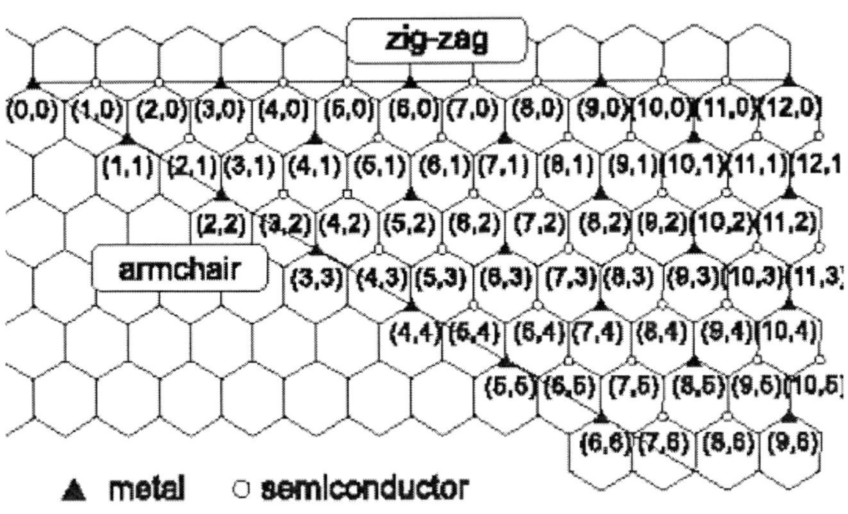

Figure 1.13. The scheme of indices (n, m) of lattice vector $\overline{R}$ tubes having semiconductor and metallic properties.

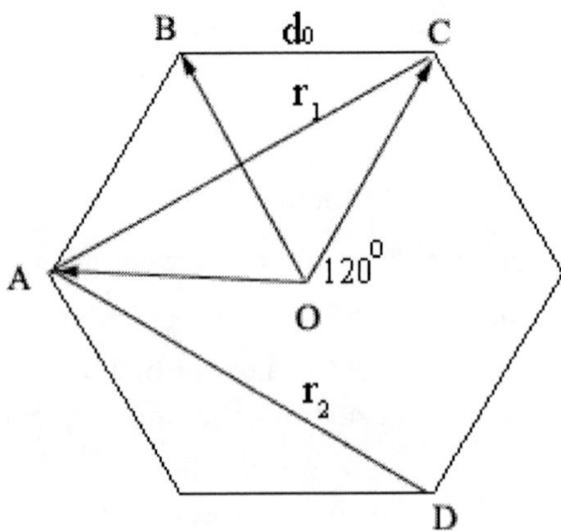

Figure 1.14. Elementary cell.

1.7. Diameter, Chirality Angle and the Mass of Single-Walled Nanotube

Indices (n, m) of single-walled nanotube chirality unambiguously determine its diameter. Therefore, the nanotubes are typically characterized by a diameter and chirality angle. Chiral angle of nanotubes is the angle between the axis of the tube and the most densely packed rows of atoms. From geometrical considerations, it is easy to deduce relations for the chiral angle and diameter of the nanotube. The angle between the basis vectors of the elementary cell (figure 1.14) $\vec{r_1}$ and $\vec{r_2}$ is equal to 60^0.

From trigonometry, $AC^2 = OA^2 + OC^2 - 2OA \cdot OC \cdot \cos 120^0$. As $OA = OC = d_0$, a $r_1 = r_2 = AC$, we have

$$r_1 = r_2 = \sqrt{3} \cdot d_0 , \tag{1.1}$$

where $d_0 = 1{,}41 \overset{0}{A} = 0{,}141\, нм$; distance between neighboring carbon atoms in the graphite plane.

Now consider the parallelogram $ABDC$ in figure 1.11a.

According to (1.1) we have

$$AB = CD = \sqrt{3}d_0 n, \qquad AC = BD = \sqrt{3}d_0 m \qquad (1.2)$$

Angle $\angle CAB = 60^0$, and $\angle ABD = 120^0$, therefore $R^2 = 3n^2 d_0^2 + 3m^2 d_0^2 - 2 \cdot 3mnd_0^2 \cos 120^0$, from which we obtain

$$R = \sqrt{3}d_0 \sqrt{n^2 + m^2 + mn}$$

Taking into account that $R = \pi \cdot d$, then to determine the diameter of the nanotube we obtain;

$$d = \frac{|\vec{R}|}{\pi} = \sqrt{3(m^2 + n^2 + mn)} \cdot \frac{d_0}{\pi} \qquad (1.3)$$

when $m = n$ we have

$$d = \frac{3nd_0}{\pi}$$

Table 1.1 shows the values of the diameters of nanotubes of different chirality.

Table 1.1.

(n, m)	d, nm	(n, m)	d, nm
(3,2)	0,334	(10,8)	1,232
(4,2)	0,417	(10,9)	1,298
(4,3)	0,480	(11,3)	1,007
(5,0)	0,394	(11,6)	1,177
(5,1)	0,439	(11,10)	1,434
(5,3)	0,552	(12,8)	1,375
(6,1)	0,517	(14,13)	1,844
(7,3)	0,701	(20,19)	2,663
(9,2)	0,801	(21,19)	2,732
(9,8)	1,161	(40,38)	5,326

Table 1.2.

CNT (n,m)	Diameter CNT, nm	Chirality
(4,0)	0,33	
(5,0)	0,39	
(6,0)	0,47	
(7,0)	0,55	
(8,0)	0,63	zigzag
(9,0)	0,70	
(10,0)	0,78	
(11,0)	0,86	
(12,0)	0,93	
(3,3)	0,40	
(4,4)	0,56	
(5,5)	0,69	armchair
(6,6)	0,81	
(7,7)	0,96	
(8,8)	1,10	
(4,1)	0,39	
(4,2)	0,43	
(7,1)	0,57	
(6,3)	0,62	chiral
(9,1)	0,75	
(10,1)	0,82	
(6,7)	0,90	

Knowing d, the chirality m and n can be found (Table 1.2). The minimal diameter of the tube is close to 0.4 nm, which corresponds to the chirality (3, 3), (5, 0), (4, 2).

We derive a formula to determine the mass of the nanotube with diameter d and length L.

The area of the elementary area ; consider a parallelogram with vertices at the centers of 4 neighboring hexagons (figure 1.15) with base $\sqrt{3}d_0$ and height $3d_0/2$ then $S_{\text{пл}} = \dfrac{3\sqrt{3}}{2}d_0^2$.

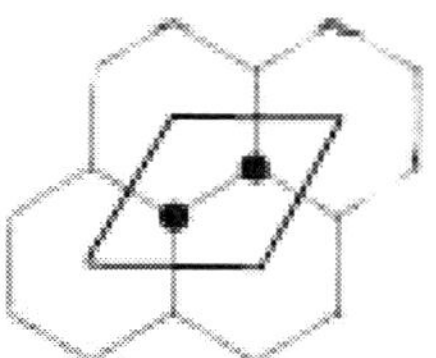

Figure 1.15. Elementary area of graphene.

The total area of the nanotube is πdL. Consequently, the number of elementary areas is equal $\pi dL / S_{пл}$.

The mass of a carbon nanotube is equal to:

$$m_T = 2m_C \frac{\pi L d}{S_{пл}} = \frac{4\sqrt{3}\pi \cdot dL}{9d_0^2} m_C \quad , \qquad (1.4)$$

where m_C =12 - mass of carbon atoms.

To determine the chiral angle θ from a right triangle AED we obtain

$$\sin\theta = \frac{DE}{R}, \quad \cos\theta = \frac{AE}{R} = \frac{\sqrt{3}nd_0 + BE}{R}$$

If we take into consideration that $\angle EDB = 30^0$, we see that $DE = \frac{3}{2}md_0$ and $BE = \frac{\sqrt{3}}{2}md_0$, consequently,

$$\sin\theta = \frac{3md_0}{2R}, \quad \cos\theta = \frac{\sqrt{3}d_0(n + m/2)}{R}$$

From these equalities we obtain the relation between the chiral indices (m,n) and angle θ:

$$\theta = arctg\left(\frac{\sqrt{3}m}{2n+m}\right) \qquad (1.5)$$

when $m = n$ we have

$$\theta = arctg\frac{\sqrt{3}}{3} \text{ OR } \theta = 30^0.$$

Chapter 2

MICRO AND NANOSCALE SYSTEMS

2.1. INTRODUCTION

Viscous forces in the fluid can lead to large dispersion flow along the axis of motion. They have a significant impact, both on the scale of individual molecules, and the scale of microflows; near the borders of the liquid-solid (beyond a few molecular layers), during the motion on a complex and heterogeneous borders.

Influence of the effect of boundary regions on the particles and fluxes have been observed experimentally in the range of molecular thicknesses up to hundreds of nanometers. If the surface has a superhydrophobic property, this range can extend to the micron thickness. *Molecular theory can predict the effect of hydrophobic surfaces in the system only up to tens of nanometers.*

Fluids, the flow of liquid or gas, have properties that vary continuously under the action of external forces. In the presence of fluid shear forces are small in magnitude, leads large changes in the relative position of the element of fluid. In contrast, changes in the relative positions of atoms in solids remain small under the action of any small external force. Termination of action of the external forces on the fluid does not necessarily lead to the restoration of its initial form.

Capillary Effects

To observe the capillary effects, one must open the nanotube, that is, to remove the upper part lids. Fortunately, this operation is quite simple.

The first study of capillary phenomena has shown that there is a relationship between the magnitude of surface tension and the possibility of its being drawn into the channel of the nanotube. It was found that the liquid penetrates into the channel of the nanotube, if its surface tension is not higher than 200 mN / m. For example concentrated nitric acid with surface tension of 43 mN/m is used to inject certain metals into the channel of a nanotube. Then annealing is conducted at 4000C for 4 h in an atmosphere of hydrogen, which leads to the recovery of the metal.

Along with the metals, carbon nanotubes can be filled with gaseous substances, such as hydrogen in molecular form. This ability is of great practical importance and can be used as a clean fuel in internal combustion engines.

Specific Electrical Resistance of Carbon Nanotubes (ρ);

The resistivity of the nanotubes can be varied within wide limits to 0.8 ohm / cm. The minimum value is lower than that of graphite. Most of the nanotubes have metallic conductivity, and the smaller shows properties of a semiconductor with a band gap of 0.1 to 0.3 eV.

The resistance of single-walled nanotube is independent of its length, because of this it is convenient to use for the connection of logic elements in microelectronic devices. The permissible current density in carbon nanotubes is much greater than in metallic wires of the same cross section and one hundred times better achievement for superconductors.

Emission Properties of Carbon Nanotubes

The results of the study of emission properties of the material (where the nanotubes were oriented perpendicular to the substrate) have been very interesting for practical use. An attained value of the emission current density is of the order of 0.5 mA / mm^2 . The value obtained is in good agreement with the Fowler-Nordheim expression.

2.2. Electro-kinetic Processes in Micro and Nanoscale Systems

The most effective and common way to control microflow substances are *electrokinetic* and *hydraulic.* At the same time the most technologically advanced and automated considered electrokinetic.

Charges transfer in mixtures occurs as a result of the directed motion of charge carriers ions. There are different mechanisms of such transfer, but usually are *convection, migration and diffusion*.

Convection is called mass transfer the macroscopic flow. *Migration* is the movement of charged particles by electrostatic fields. The velocity of the ions depends on field strength. In microfluidics the *electrokinetic process* can be divided into: *electro-osmosis, electrophoresis, streaming potential and sedimentation potential.* These processes can be qualitatively described as follows:

a) *electro-osmosis*; the movement of the fluid volume in response to the applied electric field in the channel of the electrical double layers on its wetted surfaces.
b) *Electrophoresis*; the forced motion of charged particles or molecules, in mixture with the acting electric field.
c) *Streamy potential*; the electric potential, which is distributed through a channel with charged walls, in the case when the fluid moves under the action of pressure forces. Joule electric current associated with the effect of charge transfer is flowing stream.
d) *The potential of sedimentation*; an electric potential is created when charged particles are in motion relative to a constant fluid.

In general, for the microchannel cross-section S amount of introduced probe (when entering electrokinetic method) depends on the applied voltage U, time t during which the received power, and mobility of the sample components μ:

$$Q = \frac{\mu S U t}{L} \cdot c$$

where

c ; probe concentration in the mixture,

L ; the channel length

Amount of injected substance is determined by the electrophoretic and total electroosmotic mobilities μ.

In the hydrodynamic mode of entry by the pressure difference in the channel or capillary of circular cross section, the volume of injected probe V_c:

$$V_c = \frac{4}{128} \cdot \frac{\Delta p \pi d t}{\eta L}$$

where

Δp ; pressure differential,

d ; diameter of the channel,

η ; viscosity

2.3. Continuum Hypothesis

In the simulation of processes in micron-sized systems the following basic principles are fundamentals:

1) Hypothesis of *laminar* flow (sometimes is taken for granted when it comes to microfluidics);
2) Continuum hypothesis (detection limits of applicability);
3) Laws of formation of the velocity profile, mass transfer, the distribution of electric and thermal fields;
4) Boundary conditions associated with the geometry of structural elements (walls of channels, mixers zone flows, etc.).

Since we consider the physical and chemical transport processes of matter and energy, mathematical models, have the form of systems of differential equations of second order partial derivatives. Methods for solving such equations are analytical (Fourier and its modifications, such as the method of Greenberg, Galerkin, the method of d'Alembert and the Green's functions, the Laplace operator method, etc.) or numerical (explicit or more effectively, implicit finite difference schemes).

Laminar flow; a condition in which the particle velocity in the liquid flow is not a random function of time. The small size of the microchannels (typical dimensions of 5 to 300 microns) and low surface roughness create good conditions for the establishment of laminar flow. Traditionally, the image of the nature of the flow gives the dimensionless characteristic numbers: the Reynolds number and Darcy's friction factor.

In the motion of fluids in channels the turbulent regime is rarely achieved. At the same time, the movement of gases is usually turbulent.

Although the liquids are quantized in the length scale of intermolecular distances (about 0.3 nm to 3 nm in liquids and for gases), they are assumed to be continuous in most cases. Continuum hypothesis (continuity, continuum) suggests that the macroscopic properties of fluids consisting of molecules, the same as if the fluid were completely continuous (structurally homogeneous). Physical characteristics: mass, momentum and energy associated with the volume of fluid containing a sufficiently large number of molecules must be taken as the sum of all the relevant characteristics of the molecules.

Continuum hypothesis leads to the concept of fluid particles. In contrast to the ideal of a point particle in ordinary mechanics, in fluid mechanics, particle in the fluid has a finite size.

At the atomic scale there are large fluctuations due to the molecular structure of fluids, but if we the increase the sample size, we reach a level where it is possible to obtain stable measurements. This volume of probe must contain a sufficiently large number of molecules to obtain reliable reproducible signal with small statistical fluctuations. For example, if we determine the required volume as a cube with sides of 10 nm, this volume contains some of the molecules and determines the level of fluctuations of the order of 0.5%.

The applicability of the hypothesis is based on comparison of free path length of a particle λ in a liquid with a characteristic geometric size d. The ratio of these lengths called the Knudsen number: $Kn = \lambda / d$.

When: a) $Kn < 10^{-3}$ justifies hypothesis of a continuous medium, and when b) $Kn < 10^{-1}$ allows the use of adhesion of particles to the solid walls of the channel.

Theoretical condition can also be varied: both in form $U = 0$ and in a more complex form, associated with shear stresses. The calculation of λ can be carried out as $\lambda \approx \sqrt[3]{\overline{V} / Na}$,

Where;

$\overline{V}$; molar volume,

Na ; Avogadro's number.

Under certain geometrical approximations of the particles of substance, free path length can be calculated as $\lambda \approx 1/\left(\sqrt{2}\pi r_S^2 Na\right)$, (if used instead r_S Stokes radius, as a consequence of the spherical approximation of the particle). On the other hand, for a rigid model of the molecule r_S should be replaced by the characteristic size of the particles R_g (the radius of inertia), calculated as $R_g = n_i \cdot \delta_l / \sqrt{6}$.

Here;

δ_l ; the length of a fragment of the chain (link),

n_i ; the number of links.

Of course, the continuum hypothesis is not acceptable when the system under consideration is close to the molecular scale. This happens in nanoliquid, such as liquid transport through *nano*-pores in cell membranes or artificially made nanochannels.

2.4. The Molecular Dynamics Method

In contrast to the continuum hypothesis, the essence of modeling the molecular dynamics method is as follows. We consider a large ensemble of particles which simulate atoms or molecules, i.e., all atoms are material points. It is believed that the particles interact with each other and, moreover, may be subject to external influence. Inter-atomic forces are represented in the form of the classical potential force (the gradient of the potential energy of the system).

The interaction between atoms is described by means of van der Waals forces (intermolecular forces), mathematically expressed by the Lennard-Jones potential:

$$V(r) = \frac{Ae^{-\sigma r}}{r} - \frac{C_6}{r^6}$$

where;

A and C_6 ; some coefficients depending on the structure of the atom or molecule, σ ; the smallest possible distance between the molecules.

In the case of two isolated molecules at a distance of r_0 the interaction force is zero (that is, the repulsive forces balance attractive forces). When $r > r_0$ the resultant force is the force of gravity, which increases in magnitude, reaching a maximum at $r = r_m$ and then decreases. When $r < r_0$ there is a repulsive force. Molecule in the field of these forces has potential energy $V(r)$, which is connected with the force of $f(r)$ by the differential equation

$$dV = -f(r)dr$$

At the point where $r = r_0$, $f(r) = 0$, $V(r)$ reaches an extremum (minimum).

The chart of such a potential is shown below in figure 2.1. The upper (positive) half-axis r corresponds to the repulsion of the molecules, the lower (negative) half-plane shows their attraction. We can now observe that at short distances the molecules mainly repel each other.

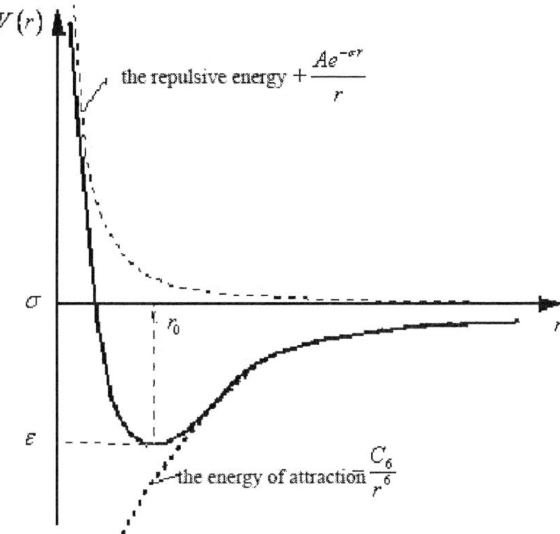

Figure 2.1. Potential energy of intermolecular interaction.

The exponential summand in the expression for the potential describing the repulsion of the molecules at small distances, often approximated as

$$\frac{Ae^{-\sigma r}}{r} \approx \frac{C_{12}}{r^{12}}$$

In this case we obtain the Lennard-Jones potential:

$$V(r) = \frac{C_{12}}{r^{12}} - \frac{C_6}{r^6} \qquad (2.1)$$

The interaction between carbon atoms is described by the potential

$$V_{CC}(r) = K(r-b)^2,$$

where;

$K = 326\ Дж / м^2$; constant tension (compression) connection,

$b = 1{,}4A$; the equilibrium length of connection,

r ; current length of the connection.

The interaction between the carbon atom and hydrogen molecule is described by the Lennard-Jones

$$V(r) = 4\varepsilon\left[\left(\frac{\sigma}{r}\right)^{12} - \left(\frac{\sigma}{r}\right)^{6}\right]$$

For all particles (figure 2.2) the equations of motion are written:

$$m\frac{d^2\overline{r_i}}{dt^2} = \overline{F}_{T-H_2}(\overline{r_i}) + \sum_{j \neq i} \overline{F}_{H_2-H_2}(\overline{r_i} - \overline{r}_j),$$

where;

$\overline{F}_{T-H_2}(\overline{r})$; Force, acting by the CNT,

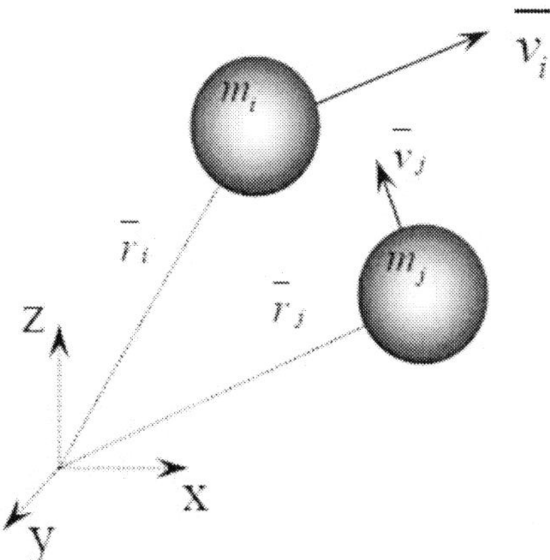

The resulting system of equations is solved numerically. However, the molecular dynamics method has limitations of applicability:

1) the de Broglie wavelength h/mv (where;
   h; Planck's constant,
   m; the mass of the particle,
   v; velocity)
2) Classical molecular dynamics can not be applied for modeling systems consisting of light atoms such as helium or hydrogen;
3) At low temperatures, quantum effects become decisive for the consideration of such systems must use quantum chemical methods;
4) The time at which we consider the behavior of the system were more than the relaxation time of the physical quantities.

Figure 2.2.

$\overline{F}_{H_2-H_2}(\overline{r}_i - \overline{r}_j)$; Force acting on the i-th molecule from the j-th molecule

The coordinates of the molecules are distributed regularly in the space. The velocities of the molecules are distributed according to the Maxwell equilibrium distribution function according to the temperature of the system:

$$f(u,v,w) = \frac{\beta^3}{\pi^{3/2}} \exp\left(-\beta^2\left(u^2+v^2+w^2\right)\right) \quad \beta = \frac{1}{\sqrt{2RT}}$$

The macroscopic flow parameters are calculated from the distribution of positions and velocities of the molecules:

$$\bar{V} = \langle \bar{v}_i \rangle = \frac{1}{n}\sum_i \bar{v}_i,$$

$$\rho = \frac{nm}{V_0}, \qquad \frac{3}{2}RT = \frac{1}{2}\left\langle \left|\bar{v}_i^{/}\right|^2 \right\rangle, \qquad \bar{v}_i^{/} = \bar{v}_i - \bar{V},$$

2.5. Van der Waals Equation. Corresponding States Law

In 1873, Van der Waals proposed an equation of state is qualitatively good description of liquid and gaseous systems. It is for one mole (one mole) is:

$$\left(p + \frac{a}{v^2}\right)(v - b) = RT \tag{2.2}$$

Note that at $p >> \frac{a}{v^2}$ and $v >> b$ this equation becomes the equation of state of ideal gas

$$pv = RT \tag{2.3}$$

Van der Waals equation can be obtained from the Clapeyron equation of Mendeleev by an amendment to the magnitude of the pressure a / v^2 and the amendment b to the volume, both constant a and b independent of T and v but dependent on the nature of the gas.

The amendment b takes into account:

1) the volume occupied by the molecules of real gas (in an ideal gas molecules are taken as material points, not occupying any volume);
2) so-called "dead space", in which can not penetrate the molecules of real gas during motion, i.e. volume of gaps between the molecules in their dense packing.

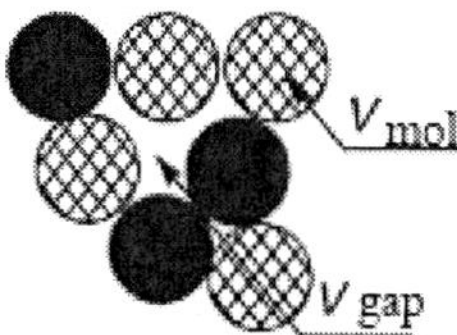

Thus, $b = v_{мол.} + v_{заз.}$ (figure 2.3). The amendment a/v^2 takes into account the interaction force between the molecules of real gases. It is the internal pressure, which is determined from the following simple considerations. Two adjacent elements of the gas will react with a force proportional to the product of the quantities of substances enclosed in these elementary volumes.

Figure 2.3. Location scheme of molecules in a real gas.

Table 2.1. The molecular weight ($[\mu] = кг / кмоль$) of some gases

gas	N	Ar	H_2	O_2	CO	CO_2	ammonia	air
μ	28	40	2	32	28	44	17	29

Therefore, the internal pressure $p_{вн}$ is proportional to the square of the concentration n:

$$p_{вн} \sim n^2 \sim \rho^2 \sim \frac{1}{v^2},$$

where ρ - the gas density.

Thus, the total pressure consists of internal and external pressures:

$$p + p_{вн} = p + \frac{a}{v^2}$$

Equation (2.3) is the most common for an ideal gas.

Under normal physical conditions ($p_0 = 0{,}1013\, МПа, t_0 = 0^0 C$) $v\mu = 22{,}4 м^3 /(кмоль \cdot K^0)$, and then from (2.3) we obtain:

$$R\mu = \frac{pv\mu}{T} = \frac{0{,}1013 \cdot 10^6 \cdot 22{,}4}{273} = 8314 \ \frac{дж}{кмоль \cdot K^0}$$

Knowing $R\mu$ we can find the gas constant for any gas with the help of the value of its molecular mass μ (Table 2.1):

$$R = \frac{R\mu}{\mu} = \frac{8314}{\mu}$$

For gas mixture with mass M state equation has the form:

$$pv = MR_{см}T = \frac{8314\ MT}{\mu_{см}} \tag{2.4}$$

where $R_{см}$ - gas constant of the mixture.

The gas mixture can be given by the mass proportions g_i, voluminous r_i or mole fractions n_i respectively, which are defined as the ratio of mass m_i, volume v_i or number of moles N_i of i gas to total mass M, volume v or number of moles N of gas mixture. Mass fraction of component is $g_i = \frac{m_i}{M}$, where $i = 1, n$. It is obvious that $M = \sum_{i=1}^{n} m_i$ and $\sum_{i=1}^{n} g_i = 1$. The volume fraction is $r_i = \frac{v_i}{v_{см}}$, where v_i - partial volume of component mixtures.

Similarly, we have $\sum_{i=1}^{n} v_i = v_{см}, \sum_{i=1}^{n} r_i = 1$.

Depending on specificity of tasks the gas constant of the mixture may be determined as follows:

$$R_{CM} = \sum_{i=1}^{n} g_i R_i \; ; \qquad R_{CM} = \frac{1}{\sum_{i=1}^{n} r_i R_i^{-1}}$$

If we know the gas constant R_{CM}, the seeming molecular weight of the mixture is equal to

$$\mu_{CM} = \frac{8314}{R_{CM}} = \frac{8314}{\sum_{i=1}^{n} g_i R_i} = 8314 \sum_{i=1}^{n} r_i R_i^{-1}$$

The pressure of the gas mixture p is equal to the sum of the partial pressures of individual components in the mixture p_i:

$$p = \sum_{i=1}^{n} p_i \tag{2.5}$$

Partial pressure p_i - pressure that has gas, if it is one at the same temperature fills the whole volume of the mixture ($p_i v_{CM} = RT$).

With various methods of setting the gas mixture partial pressures

$$p_i = p r_i \; ; \qquad p_i = \frac{p g_i \mu_{CM}}{\mu_i} \tag{2.6}$$

From the expression (2.6) we see that for the calculation of the partial pressures p_i necessary to know the pressure of the gas mixture, the volume or mass fraction i of the gas component, as well as the molecular weight of the gas mixture μ and the molecular weight of i of gas μ_i.

The relationship between mass and volume fractions are written as follows:

$$g_i = \frac{m_i}{m_{CM}} = \frac{\rho_i v_i}{\rho_{CM} v_{CM}} = \frac{R_{CM}}{R_i} r_i = \frac{\mu_i}{\mu_{CM}} r_i$$

We rewrite (2.2) as

$$v^3 - \left(b + \frac{RT}{p}\right)v^2 + \frac{a}{p}v - \frac{ab}{p} = 0 \quad (2.7)$$

When $p = p_k$ and $T = T_k$, where p_k and T_k - critical pressure and temperature, all three roots of (2.7) are equal to the critical volume v_k

$$v^3 - \left(b + \frac{RT_k}{P_k}\right)v^2 + \frac{a}{p_k}v - \frac{ab}{p_k} = 0 \quad (2.8)$$

Because the $v_1 = v_2 = v_3 = v_k$, then equation (2.8) must be identical to the equation

$$(v - v_1)(v - v_2)(v - v_3) = (v - v_k)^3 = v^3 - 3v^2 v_k + 3v v_k^2 - v_k^3 = 0 \quad (2.9)$$

Comparing the coefficients at the equal powers of v in both equations leads to the equalities

$$b + \frac{RT_k}{p_k} = 3v_k \;;\; \frac{a}{p_k} = 3v_k^2 \;;\; \frac{ab}{p_k} = v_k^3 \quad (2.10)$$

Hence

$$a = 3v_k^2 p_k; \qquad b = \frac{v_k}{3} \quad (2.11)$$

Considering (2.10) as equations for the unknowns p_k, v_k, T_k, we obtain

$$p_k = \frac{a}{27b^2}; \qquad v_k = 3b; \qquad T_k = \frac{8a}{27bR}. \qquad (2.12)$$

From (2.10) and (2.11) or (2.12) we can find the relation

$$\frac{RT_k}{p_k v_k} = \frac{8}{3} \qquad (2.13)$$

Instead of the variables p, v, T let's introduce the relationship of these variables to their critical values (leaden dimensionless parameters)

$$\pi = \frac{p}{p_k}; \qquad \omega = \frac{v}{V_k}; \qquad \tau = \frac{T}{T_k}. \qquad (2.14)$$

Substituting (2.12) and (2.14) in (2.7) and using (2.13), we obtain

$$\left(\pi p_k + \frac{3v_k^2 p_k}{\omega^2 v_k^2}\right)\left(\omega v_k - \frac{v_k}{3}\right) = RT_k\tau,$$

$$\left(\pi + \frac{3}{\omega^2}\right)(3\omega - 1) = 3\frac{RT_k}{p_k v_k}\tau,$$

$$\left(\pi + \frac{3}{\omega^2}\right)(3\omega - 1) = 8\tau \qquad (2.15)$$

In (2.15) a and b are not permanent, depending on the nature of the gas. That is, if the units of measurement of pressure, volume and temperature are used as their critical values (use the leaden parameters), the equation of state is the same for all substances.

This condition is called the *law of corresponding states.*

Chapter 3

RHEOLOGICAL PROPERTIES AND STRUCTURE OF THE LIQUID IN NANOTUBES

3.1. SLIPPAGE OF THE FLUID PARTICLES NEAR THE WALL

According to the Navier boundary condition the velocity slip is proportional to fluid velocity gradient at the wall:

$$v\big|_{y=0} = L_S \, dv / dy\big|_{y=0} \tag{3.1}$$

Here and in figure 3.1;

L_S represents the "slip length" and has a dimension of length.

Because of the slippage, the average velocity in the channel $\langle v_{pdf} \rangle$ increases.

In a rectangular channel (of width >> height h and viscosity of the fluid η) due to an applied pressure gradient of dp/dx:

$$\langle v_{pdf} \rangle = \frac{h^2}{12\eta}\left(-\frac{dp}{dx}\right)\left(1+\frac{6L_S}{h}\right) \tag{3.2}$$

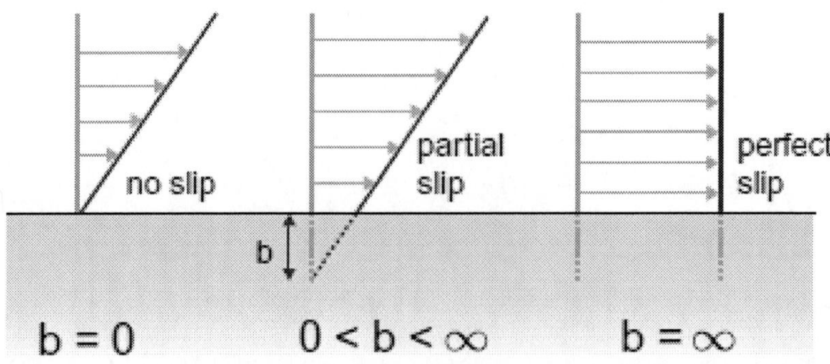

Figure 3.1. Three cases of slip flow (with slip length *b)*.

The results of molecular dynamics simulation for nanosystems with liquid, show that a large slippage lengths (of the order of microns) should occur in the carbon nanotubes of nanometer diameter and, consequently, can increase the flow rate by three orders of ($6L_S / h > 1000$). Thus, the flow with slippage is becoming more and more important for hydrodynamic systems of small size.

The results of molecular dynamics simulation of unsteady flow of mixtures of water - water vapor, water and nitrogen in a carbon nanotube are reported previously. Based on these studies, a flow of water through carbon nanotubes with different diameters at temperature of $300^0 K$ was considered.

Carbon nanotubes have been considered "zigzag" with chiral vectors (20, 20), (30, 30) and (40, 40), corresponding to pipe diameters or 2,712, 4,068 and 5,424 nm, respectively.

The value of the flow rate and the system pressure, which varies in the range of 600-800 bars, were high enough to ensure complete filling of the tubes. This pressure was achieved by the total number of water molecules 736, 904, and 1694, respectively.

The effects of slippage of various liquids on the surface of the nanotube were studied in detail.

The length of slip, calculated using the current flow velocity profiles of liquid, shown in figure 3.2, were 11, 13, and 15 nm for the pipes of 2,712, 4,068 and 5,424 nm respectively. The dotted line marked by theoretical modeling data. The vertical lines indicate the position of the surface of carbon nanotubes.

It was found out that as the diameter decreases, the speed of slippage of particles on the wall of nanotube also decreases. The report attributes this to the increase of the surface friction.

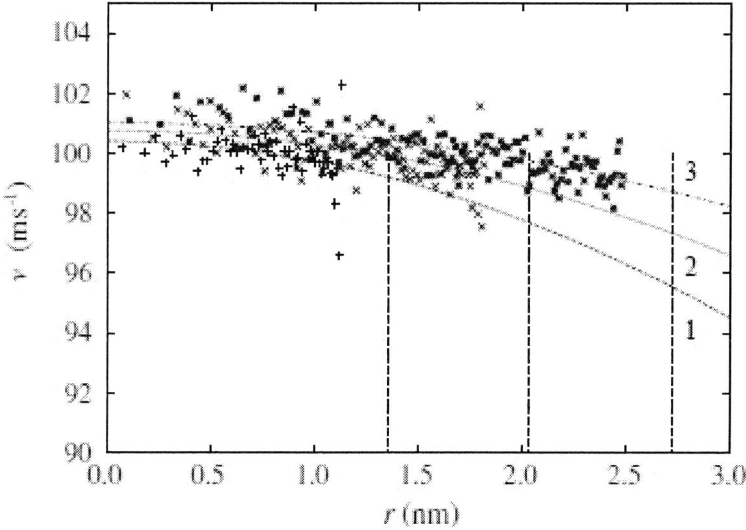

Figure 3.2. Time average streaming velocity profiles of water in a carbon nanotubes of different diameter: 2.712 nm, curve 1; 4.068 nm, curve 2; 5.424 nm, curve 3.

Experiments with various pressure drops in nanotubes demonstrated slippage of fluid in micro-and nanosystems. The most remarkable were the two recent experiments, which were conducted to improve the flow characteristics of carbon nanotubes with the diameters of 2 and 7 nm, respectively.

In the membranes in which the carbon nanotubes were arranged in parallel, there was a slip of the liquid in the micrometre range. This led to a significant increase in flow rate, up to three and four orders of magnitude.

In the experiments for the water moving in microchannels on smooth hydrophobic surfaces, there are sliding at about 20 nm. If the wall of the channel is not smooth but twisty or rough (and hydrophobic) such a structure would lead to an accumulation of air in the cavities and become superhydrophobic (with contact angle greater than 160^0). It is believed that this leads to creation of contiguous areas with high and low slippage, which can be described as "effective slip length". This effective length of the slip occurring on the rough surface can be several tens of microns, which was confirmed experimentally.

It should be noted that for practical use of advantages of nanotubes with slippage is necessary to solve many more problems. Scoentists have already shown that hydrophobic surfaces tend to form bubbles. On the other hand, the surfaces used by most researchers were rough, but the use of smooth surfaces could generally reduce the formation of bubbles.

Another possible problem is filling of the hydrophobic systems with liquid. Filling of micron size hydrophobic capillaries is not a big problem, because pressure of less than 1 atm is sufficient. Capillary pressure, however, is inversely proportional to the diameter of the channel, and filling for nanochannels can be very difficult.

3.2. The Density of the Liquid Layer Near a Wall of Carbon Nanotube

Scientists showed radial density profiles of oxygen averaged in time and hydrogen atoms in the "zigzag" carbon nanotube with chiral vector (20, 20) and a radius $R = 1{,}356$ nm (figure 3.3). The distribution of molecules in the area near the wall of the carbon nanotube indicated a high density layer near the wall of the carbon nanotube. Such a pattern indicates the presence of structural heterogeneity of the liquid in the flow of the nanotube. In figure (3.3) $\rho^*(r) = \rho(r)/\rho(0)$, whereas 2,712 nm diameter pipe is completely filled with water molecules at $300^0 K$. The overall density is $\rho(0) = 1000\ кг/м^3$. The arrows denote the location of distinguishable layers of the water molecules and the vertical line is the position of the CNT wall.

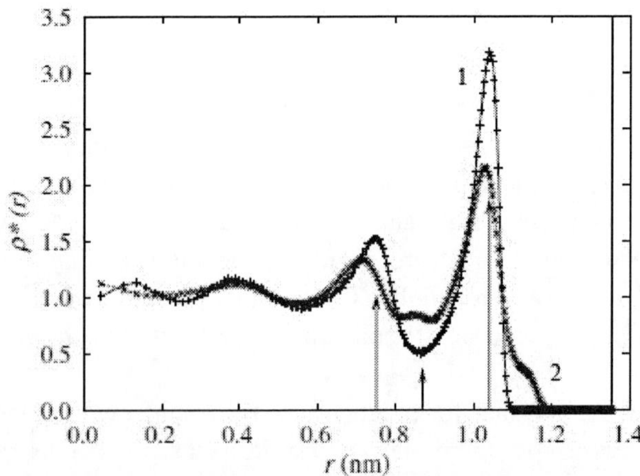

Figure 3.3. Radial density profiles of oxygen (curve 1) and hydrogen (curve 2) atoms.

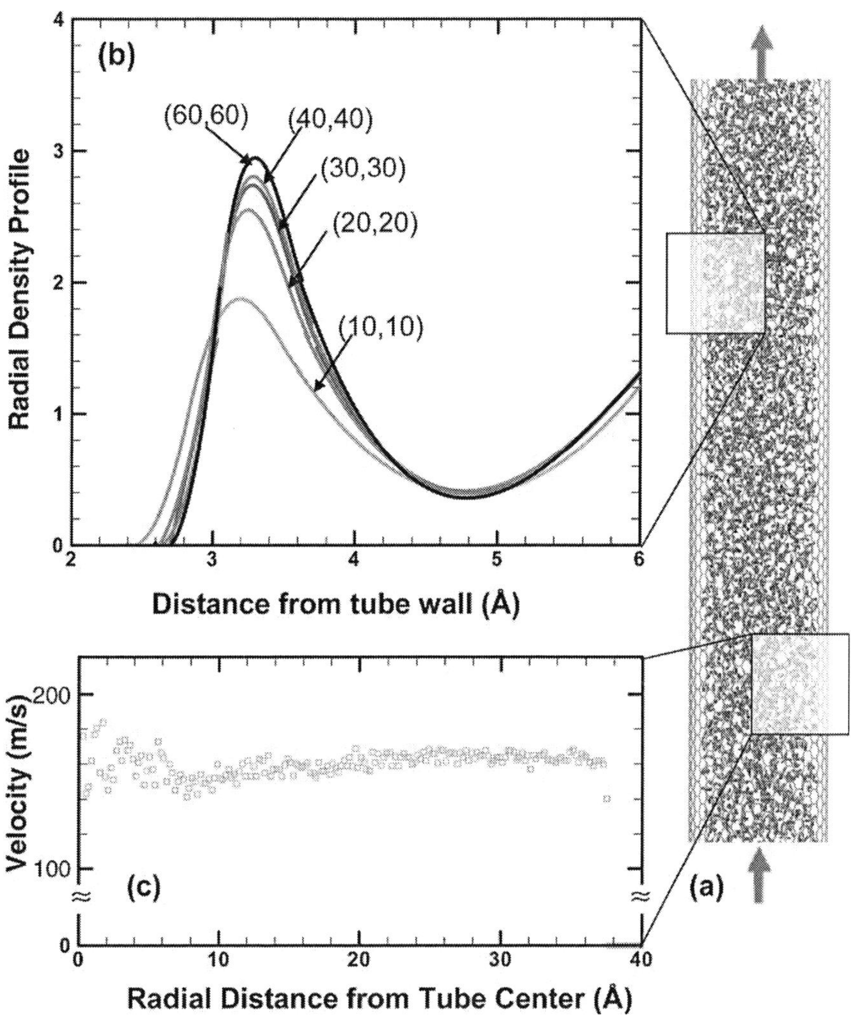

Figure 3.4. initial structure and movement of water molecules: (a) Schematic of the initial structure and transport of water molecules in a model CNT. (b) The radial density profile (RDP) of water molecules inside CNTs with different radii. (c) The representative radial velocity profile (RVP) of water molecules inside a (60,60) nanotube (Xi Chen et al. 2008).

The distribution of molecules in the region of $0{,}95 \leq r \leq R$ nm indicates a high density layer near the wall carbon nanotube. Such a pattern indicates the presence of structural heterogeneity of the liquid in the flow of the nanotube.

Figure 3.4 shows the scheme of the initial structure and movement of water molecules in;

(a) the model carbon nanotube;
(b) the radial density profile of water molecules inside nanotubes with different radii and;
(c) the velocity profile of water molecules in a nanotube chirality (60,60).

It should be noted that the available area for molecules of the liquid is less than the area bounded by a solid wall, primarily due to the van der Waals interactions.

3.3. THE EFFECTIVE VISCOSITY OF THE LIQUID IN A NANOTUBE

There is a significant increase in the effective viscosity of the fluid in the nano volumes compared to its macroscopic value. However, the effective viscosity of the liquid in a nanotube depends on the diameter of the nanotube.

The effective viscosity of the liquid in a nanotube is defined as follows.

A conformity nanotubes filled with liquid, containing crystallites with the same size tube filled with liquid, can be considered as a homogeneous medium (i.e. without considering the crystallite structure), in which the same pressure drop and flow rate of Poiseuille flow is realized. The viscosity of a homogeneous fluid, which ensures the coincidence of these parameters, called the effective viscosity of the flow in the nanotube.

While flowing in the narrow channels of width less than 2 nm, water behaves like a viscous liquid. In the vertical direction, water behaves as a rigid body, and in a horizontal direction it maintains its fluidity.

It is known that at large distances the van der Waals interaction has a magnetic tendency and occurs between any molecules like polar as well as nonpolar. At small distances it is compensated by repulsion of electron shells. Van der Waals interaction decreases rapidly with distance. Mutual convergence of the particles under the influence magnetic forces continues until these forces are balanced with the increasing forces of repulsion.

Knowing the deceleration of the flow (figure 3.4) of water a, the effective shear stress between the wall of the pipe length l and water molecules can be calculated by;

$$\tau = Nma/(2\pi Rl) \tag{3.3}$$

Here, the shear stress is a function of tube radius and flow velocity $\bar{v}$, and m is mass of water molecules. The average speed is related to volumetric flow $\bar{v} = Q/(\pi R^2)$.

Denoting n_0 the density of water molecules number, we can calculate the shear stress in the form of:

$$\tau\big|_{r=R} = n_0 mRa/2 \tag{3.4}$$

Figure 3.5 shows the results of calculations concerning the influence of the size of the tube R_0 on the effective viscosity (squares) and shear stress τ (triangles), when the flow rate is approximately 165 m/sec.

According to classical mechanics of liquid flow at different pressure drops Δp along the tube length l is given by Poiseuille formula;

$$Q_P = \frac{\pi R^4 \Delta p}{8\eta l}, \tag{3.5}$$

Therefore

$$\tau = \frac{\Delta p R}{2l} \tag{3.6}$$

and the effective viscosity of the fluid can be estimated as $\eta = \tau \cdot R/(4\bar{v})$.

The change in the value of shear stress directly causes the dependence of the effective viscosity of the fluid from the pipe size and flow rate. In this case the effective viscosity of the transported fluid can be determined from (3.5) and (3.6) as

$$\eta = \frac{\tau \cdot R}{4\bar{v}} \tag{3.7}$$

It should be noted that the magnitude of the shear stress τ is relatively small in the range of pipe sizes considered. This indicates that the surface of carbon nanotubes is very smooth and the water molecules can easily slide through it.

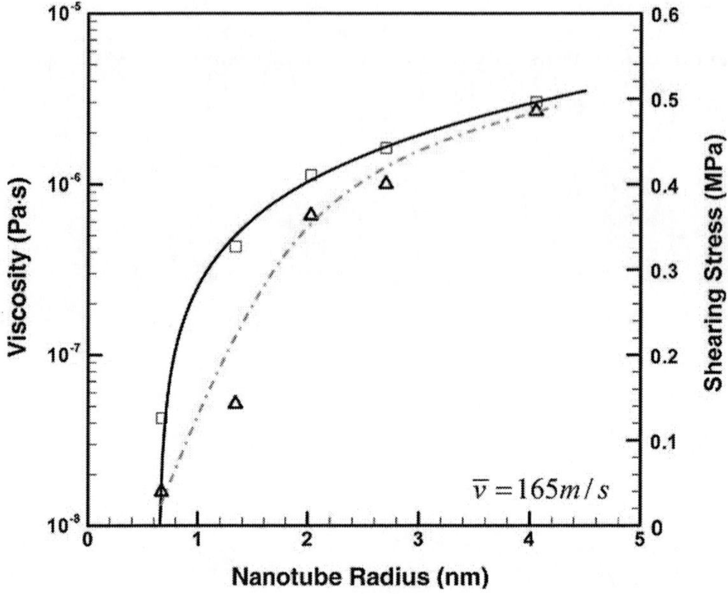

Figure 3.5. Size effect of shearing stress (triangle) and viscosity (square).

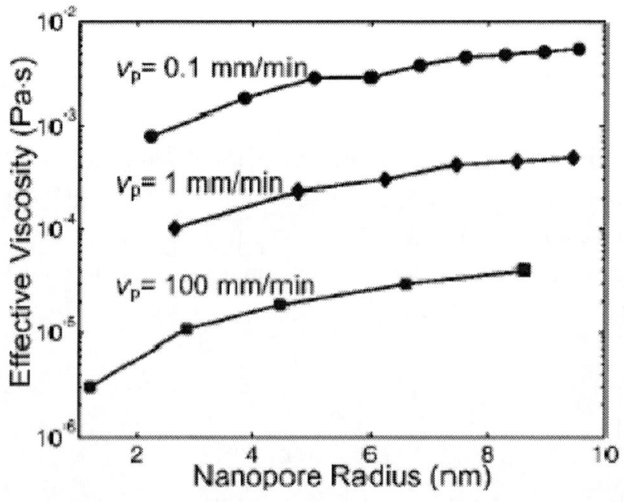

Figure 3.6. Effective viscosity as a function of the nanopore radius and the loading rate.

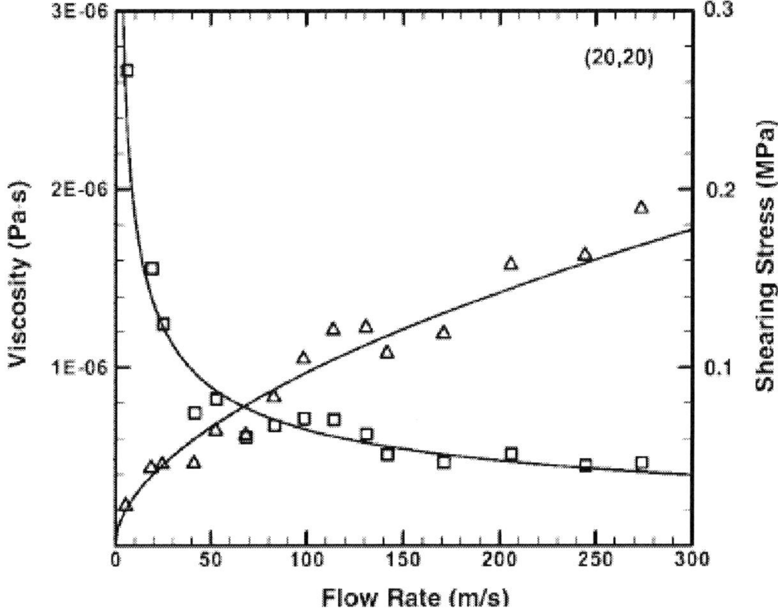

Figure 3.7. Flow rate effect of shearing stress (triangle) and viscosity (square).

In fact, shear stress is primarily due to van der Waals interaction between the solid wall and the water molecules. It is noted that the characteristic distance between the near-wall layer of fluid and pipe wall depends on the equilibrium distance between atoms O and C and the distribution of the atoms of the solid wall and bend of the pipe.

From Figure 3.5 the effective viscosity η increases by two orders of magnitude when R_0 changing from 0.67 to 5.4 nm.

According to equations (3.5) - (3.7), the effective viscosity can be calculated as $\eta = \dfrac{\pi R^4 \Delta p}{8QL}$. The results of calculations are shown in figure 3.6.

The dependence of the shear stress on the flow rate is illustrated in Figure 3.7. For the tube (20,20) τ increases with $\bar{v}$. The growth rate slowed down at higher values $\bar{v}$.

At high speeds $\bar{v}$, while water molecules are moving along the surface of the pipe, the liquid molecules do not have enough time to fully adjust their positions to minimize the free energy of the system. Therefore, the distance between adjacent carbon atoms and water molecules may be less than the equilibrium van der Waals distances. This leads to an increase in van der Waals forces of repulsion and leads to higher shear stress.

It should be noted that even though the equation for viscosity is based on the theory of the continuum, it can be extended to a complex flow to determine the effective viscosity of the nanotube.

Figure 3.7 shows dependence of η on $\bar{v}$ inside of the nanotube (20, 20). It is shown that η decreases sharply with increasing flow rate and begins to approach a definite value when $\bar{v}$>150 m/sec. For the current pipe size and flow rate ranges of $\eta \sim 1/\sqrt{\bar{v}}$, this trend is because of τ - $\bar{v}$, contained in figure 3.7. According to figure 3.6 high-speed effects are negligible.

One can easily see that the dependence of viscosity on the size and speed is consistent qualitatively with the results of molecular dynamic simulations. In all studied cases, the viscosity is much smaller than its macroscopic analogy. As the radius of the pores varies from about 1 nm to 10 nm, then the value of the effective viscosity increases by an order of magnitude. A more significant change occurs when increasing the speed of 0,1 mm / min up to 100 mm / min. This results in a change in the value of viscosity η, respectively, by 3-4 orders. The discrepancy between simulation and test data can be associated with differences in the structure of the nanopores and liquid phase.

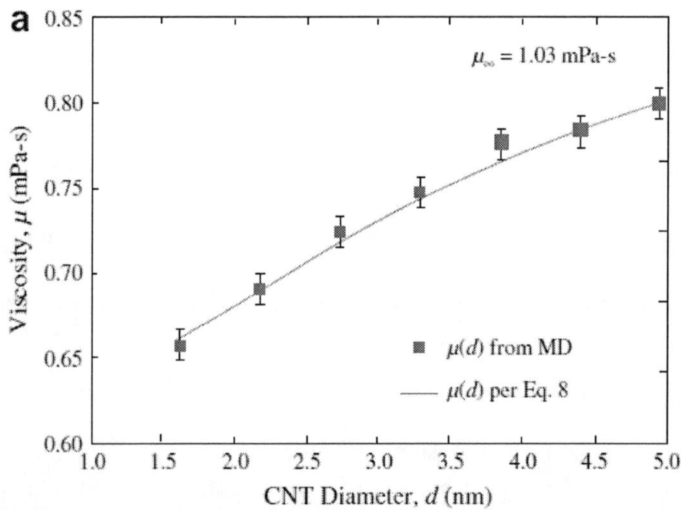

Figure 3.8. Variation of water viscosity with CNT diameter.

Figure 3.8 shows the viscosity dependence of water (calculated by the method of DM, the diameter of the CNT). The viscosity of water, as shown in the figure, increases monotonically with increasing diameter of the CNT.

3.4. The Release of Energy due to the Collapse of the Nanotube

Scentists theoretically predicted the existence of a "domino effect" in single-walled carbon nanotube.

Squashing can occur at one end by two rigid movements of narrow graphene planes (about 0.8 nm in width and 8.5 nm in length). This can rapidly (at a rate exceeding 1 km/s) release its stored energy by collapsing along its length like a row of dominoes. The effect resembles a tube of toothpaste squeezing itself (figure 3.9).

The structure of a single-walled carbon nanotube has two possible stable states: circular or collapsed. Scientists realized that for nanotubes wider than 3.5 nanometres, the circular state stores more potential energy than the collapsed state as a result of van der Waal's forces. He performed molecular dynamics simulations to find out what would happen if one end of a nanotube was rapidly collapsed by clamping it between two graphene bars.

This phenomenon occurs with the release of energy, and thus allows for the first time to talk about carbon nanotubes as energy sources. This effect can also be used as an accelerator of molecules.

The tube collapses at the same time not over its entire length, and sequentially, one after the other carbon ring, starting from the end, which is tightened (figure 3.9.). It happens just like a domino collapses, arranged in a row (this is known as the "domino effect"). The role of bone dominoes performing here as ring of carbon atoms forming the nanotube, and the nature of this phenomenon is quite different.

Recent studies show that for nanotubes with diameters ranging from 2 to 6 nm, there are two stable equilibrium states;

- Cylindrical (tube no collapses) and;
- Compressed (imploded tube) with different values of potential energy (the difference between which and can be used as an energy source).

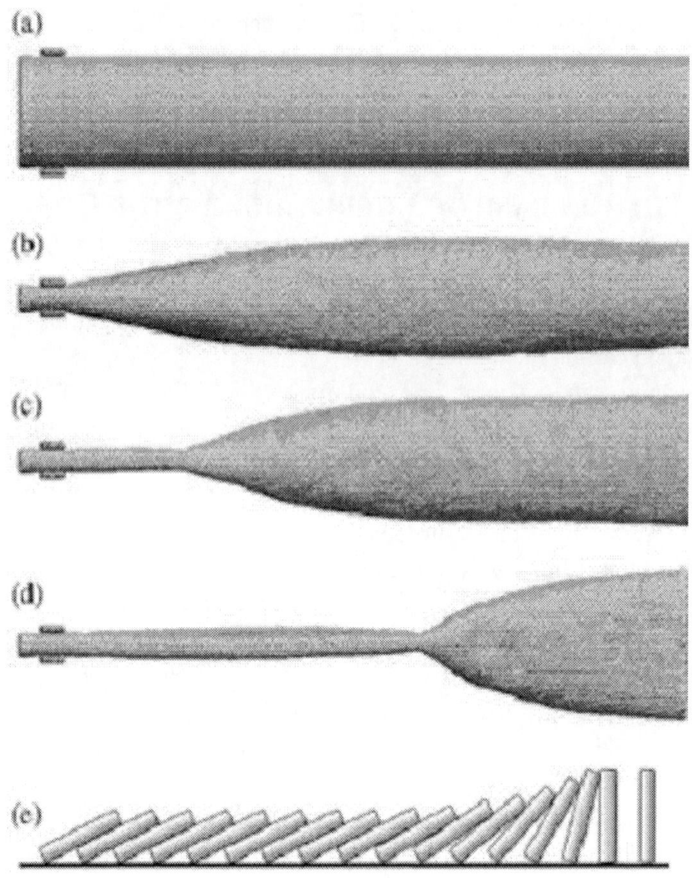

Figure 3.9. "Domino effect" in a carbon nanotube. (a) The initial form of carbon nanotubes is cylindrical. (b) One end of the tube is squeezed. (c), (d) Propagation of domino waves; the configuration of the nanotube 15 and 25 picoseconds after the compression of its end. (e) Schematic illustration of the "domino effect" under the influence of gravity.

The switching between these two states with the subsequent release of energy occurs in the form of waves arising domino effect. The scientists have shown that such switching is carried out in carbon nanotubes with diameters of 2 nm.

A theoretical study of the "domino effect" was conducted by using a special method of classical molecular dynamics, in which the interaction between carbon atoms was described by van der Waals forces.

The main reason for the observed effect, is the potential energy of the van der Waals interactions (which "collapses" the nanotube) and the energy of elastic deformation (which seeks to preserve the geometry of carbon atoms). This eventually leads to a bistable (collapsed and no collapsed) configuration carbon nanotube.

Thus, “domino effect” wave can be produced in a carbon nanotube with a relatively large diameter (more than 3.5 nm), because only in such a system the potential energy of collapsing structures may be less than the potential energy of the "normal" nanotube. In other words, the cylindrical structure and collapsing nanotubes with large diameters are, respectively, of the meta-stable and stable states.

The potential energy of a carbon nanotube with a propagating wave at the “domino effect” represented in figure 3.10a.

This figure shows three sections;

The first (from 0 ps to 10 ps) are composed of elastic strain energy, which appears due to changes in the curvature of the walls of the nanotubes in the process of collapsing.

The second region (from 10 ps to 35 ps) corresponds to the "domino effect".

Finally, the third segment (from 35 ps to 45 ps) corresponds to the process is ended "domino".

The spread "domino effect" waves are a process that goes with the release of energy (about 0.01 eV per atom of carbon). This is certainly not comparable in any way with the degree of energy yield in nuclear reactions, but the fact of power generation carbon nanotube is obvious.

Calculations show that the wave of dominoes in a tube with diameter of 4-5 nm is about 1 km / s (as seen from the figure 3.10b) and non-linear manner depending on its geometry (the diameter and chirality). The maximum effect should be observed in the tube with a diameter slightly less than 4.5 nm (considering carbon rings to collapse at a speed of 1.28 km / sec). The theoretical dependence shows the blue solid line. And now an example of how energy is released in such a system with a "domino effect" can be used in nano-devices. Scientists offer an original way to use sort of "nanogun" (figure 3.11a). Imagine that at our disposal is a carbon nanotube with chirality (55.0) and related to the observation of the dominoes in diameter. Put inside a nanotube fullerene C_{60}. With a little imagination, this can be considered a carbon nanotube as the gun trunk, and the molecule (as its shell).

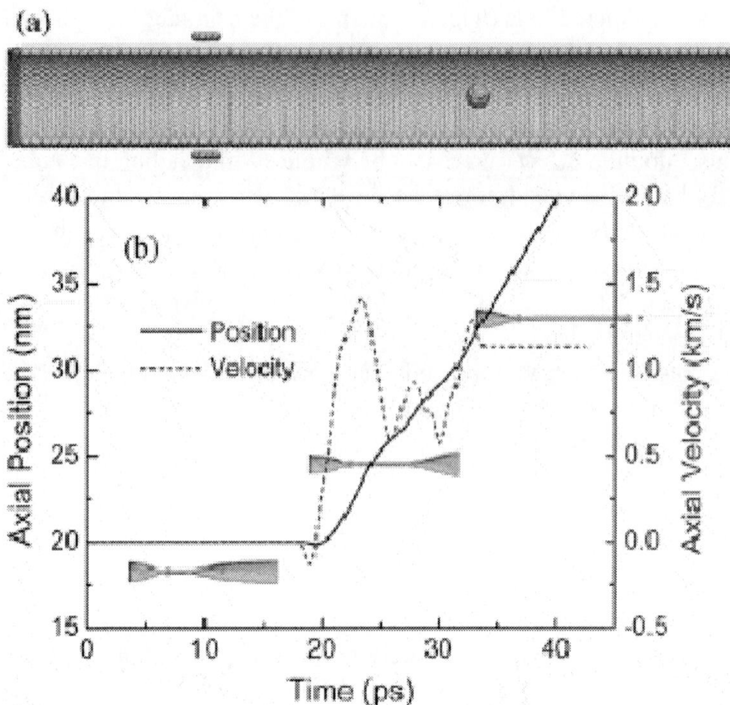

Figure 3.11. Nano cannon scheme acting on the basis of "domino effect" in the incision. (a) Inside a carbon nanotube (55.0) is the fullerene molecule C_{60}. (b) The initial position and velocity of the departure of the "core" (a fullerene molecule), depending on the time. The highest rate of emission of C_{60} (1.13 km /s) comparable to the velocity of the domino wave.

The question is what is the speed of the "core"? Based on the initial position of the fullerene molecule on departure of a nanotube, it can reach speeds close to the velocity of "domino effect" waves which is about 1 km/s (figure 3.11b). Interestingly, this speed is reached by the "core" for just 2 picoseconds and at a distance of 1 nm. It is easy to calculate that the observed acceleration is of great value $0{,}5 \cdot 10^{15}\, м / c^2$.

Chapter 4

FLUID FLOW IN NANOTUBES

4.1. INTRODUCTION

Scientists from the University of Wisconsin-Madison (USA) managed to prove that the laws of friction for the nanostructures do not differ from the classical laws.

The friction of surface against the surface in the absence of the interlayer between the liquid materials (so-called dry friction) is created by irregularities in the given surfaces that rub one another, as well as the interaction forces between the particles that make up the surface.

As part of their study, the researchers built a computer model that calculates the friction force between nano-surfaces (figure 4.1). In the model, these surfaces were presented simply as a set of molecules for which forces of intermolecular interactions were calculated.

As a result, scientists were able to establish that the friction force is directly proportional to the number of interacting particles. The researchers propose to consider this quantity by analog of so-called true macroscopic contact area. It is known that the friction force is directly proportional to this area (it should not be confused with common area of the contact surfaces of the bodies).

In addition, the researchers were able to show that the friction surface of the nano-surfaces can be considered within the framework of the classical theories of friction of non-smooth surfaces.

A literature review shows that nowdays molecular dynamics and mechanics of the continuum in are the main methods of research of fluid flow in nanotubes.

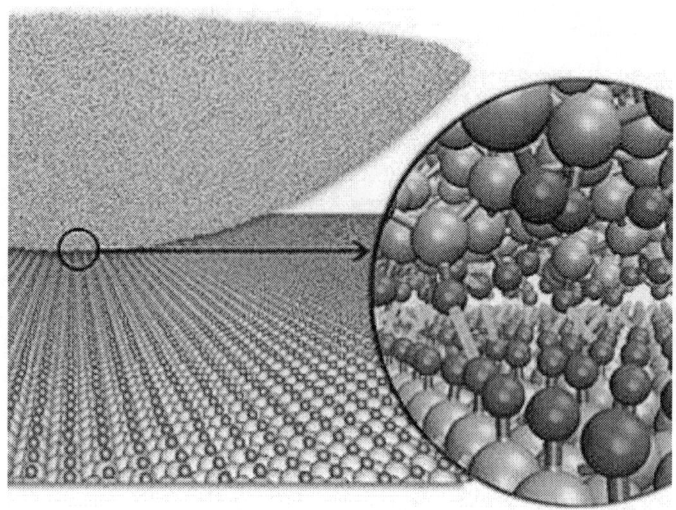

Figure 4.1. Computer model of friction at the nanoscale The right shows the surfaces of interacting particles (Friction laws at the nanoscale. Nature, 26.02.2009).

Although the method of molecular dynamics simulations is effective, it at the same time requires enormous computing time especially for large systems. Therefore, simulation of large systems is more reasonable to carry out nowadays by the method of continuum mechanics (Yoon J. et al., 2003), (Wang CY et al., 2005), (Natsuki T. et al., 2006), (Yoon J. et al., 2003), (Yoon J. et al., 2004), (Natsuki T. et al., 2005), (Wang Q. and Varadan VK, 2006).

In the work of (Morten Bo Lindholm Mikkelsen et al., 2005) the fluid flow in the channel is considered in the framework of the continuum hypothesis. The Navier-Stokes equation was used and the velocity profile was determined for Poiseuille flow.

In the work of (Thomas John A. and McGaughey Alan JH, 2008) the water flow by means of pressure differential through the carbon nanotubes with diameters ranging from 1.66 to 4.99 nm is researched using molecular dynamics simulation study. For each nanotube the value enhancement predicted by the theory of liquid flow in the carbon nanotubes is calculated. This formula is defined as a ratio of the observed flow in the experiments to the theoretical values without considering slippage on the model of Hagen-Poiseuille. The calculations showed that the enhancement decreases with increasing diameter of the nanotube.

Important conclusion of the (Thomas John A. and McGaughey Alan JH, 2008) is that by constructing a functional dependence of the viscosity of the water and length of the slippage on the diameter of carbon nanotubes, the experimental results in the context of continuum fluid mechanics can easily be described. The aforementioned is true even for carbon nanotubes with diameters of less than 1,66 nm.

The theoretical calculations use the following formula for the steady velocity profile of the viscosity η of the fluid particles in the CNT under pressure gradient $\partial p / \partial z$:

$$v(r) = \frac{R^2}{4\eta}\left[1 - \frac{r^2}{R^2} + \frac{2L_S}{R}\right]\frac{\partial p}{\partial z} \tag{4.1}$$

The length of the slip, which expresses the speed heterogeneity at the boundary of the solid wall and fluid is defined as in (Joseph P. Et al. 2005), (Barrat J.-L. and Chiaruttini F., 2003), (Sokhan VP Et al., 2002):

$$L_S = \left.\frac{v(r)}{dv/dr}\right|_{r=R} \tag{4.2}$$

Then the volumetric flow rate, taking into account the slip Q_S is defined as:

$$Q_S = \int_0^R 2\pi r \cdot v(r) dr = \frac{\pi\left[(d/2)^4 + 4(d/2)^3 \cdot L_S\right]}{8\eta} \cdot \frac{\partial p}{\partial z} \tag{4.3}$$

Equation (4.3) is a modified Hagen-Poiseuille equation, taking into account slippage. In the absence of slip $L_S = 0$ (4.3) coincides with the Hagen-Poiseuille flow (3.5) for the volumetric flow rate without slip Q_P. In the works (Holt et al., 2006 and Majumder et al., 1999) the parameter enhancement flow ε is introduced. It is defined as the ratio of the calculated volumetric flow rate of slippage to Q_P (calculated using the effective viscosity and the diameter of the CNT). If the measured flux is modeled using equation (4.3), the degree of enhancement takes the form:

$$\varepsilon = \frac{Q_S}{Q_P} = \left[1 + 8\frac{L_S(d)}{d}\right]\frac{\eta_\infty}{\eta(d)} \tag{4.4}$$

where;

$d = 2R$; diameter of CNT,

η_∞ ; viscosity of water,

$L_S(d)$;CNT slip length depending on the diameter,

$\eta(d)$ -;the viscosity of water inside CNTs depending on the diameter.

-If $\eta(d)$ finds to be equal to η_∞, then the influence of the effect of slip on ε is significant,

- if $L_S(d) \geq d$. If $L_S(d) << d$ and $\eta(d) = \eta_\infty$, then there will be no significant difference compared to the Hagen-Poiseuille flow with no slip.

Table 4.1 shows the experimentally measured values of the enhancement water flow. Enhancement flow factor and the length of the slip were calculated using the equations given above.

Figure 4.2 depicts the change in viscosity of the water and the length of the slip in diameter. As can be seen from the figure, the dependence of slip length to the diameter of the nanotube is well described by the empirical relation

$$L_S(d) = L_{S,\infty} + \frac{C}{d^3} \tag{4.5}$$

where;

$L_{S,\infty}$ =30 nm ; slip length on a plane sheet of graphene,

C ; const.

Table 4.1.

Nanosystems	Diameter (nm)	Enhancement, ε	slip length, L_S, (nm)
carbon nanotubes	300-500	1	0
	44	22-34	113-177
carbon nanotubes	7	$10^4 - 10^5$	3900-6800
	1,6	560-9600	140-1400

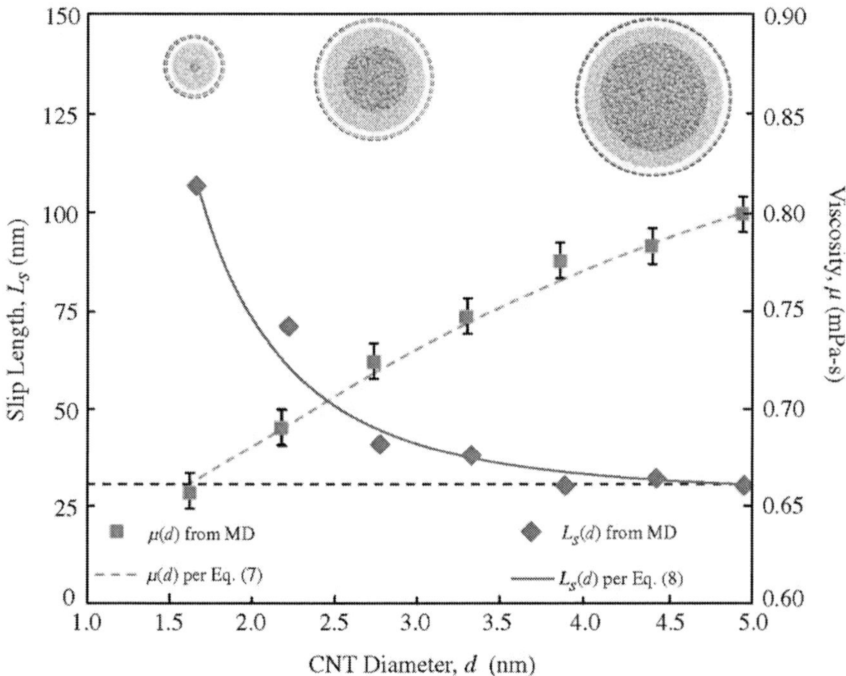

Figure 4.2. Variation of viscosity and slip length with CNT diameter (Thomas John A. and McGaughey Alan J. H., 2008).

Figure 4.3 shows dependence of the enhancement of the flow rate ε on the diameter for all seven CNTs.

There are three important features in the results;

- First, the enhancement of the flow decreases with increasing diameter of the CNT. -Second, with increasing diameter, the value tends to the theoretical value of (4.4) and (4.5) with a slip $L_{s\infty}$ = 30 nm and the effective viscosity $\mu(d) = \mu_\infty$. The dotted line showed the curve of 15% in the second error in the theoretical data of viscosity and slip length.
- Third, the change ε in diameter of CNTs can not be explained only by the slip length.

To determine the dependence the volumetric flow of water from the pressure gradient along the axis of single-walled nanotube with the radii of 1.66, 2.22, 2.77, 3.33, 3.88, 4.44 and 4.99 nm in (Thomas John A. and

McGaughey Alan JH, 2008) the method of molecular modeling was used. Snapshot of the water-CNT is shown in figure 4.4.

Figure 4.4 shows the results of calculations to determine the pressure gradient along the axis of the nanotube with the diameter of 2.77 nm and a length of 20 nm. Change of the density of the liquid in the cross sections was less than 1%.

Figure 4.5 shows the dependence of the volumetric flow rate from the pressure gradient for all seven CNTs. The flow rate ranged from 3-14 m/sec. In the range considered here the pressure gradient $(0-3)\cdot 10^{9}\, atm/m$ Q ($pl/sek = 10^{-15} m^{3}/sek$) is directly proportional to $\partial p / \partial z$. Coordinates of chirality for each CNT are indicated in the figure legend. The linearity of the relations between flow and pressure gradient confirms the validity of calculations of the formula (4.3).

Figure 4.6 shows the profile of the radial velocity of water particles in the CNT with diameter 2.77 nm. The vertical dotted line at 1.38 nm marked surface of the CNT. It is seen that the velocity profile is close to a parabolic shape.

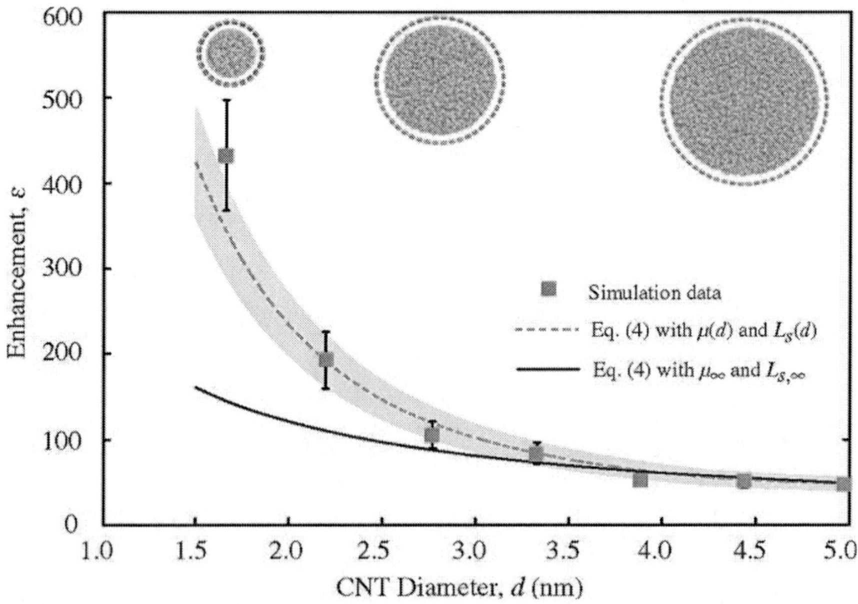

Figure 4.3. Flow enhancement as predicted from MD simulations (Thomas John A. and McGaughey Alan J. H., 2008).

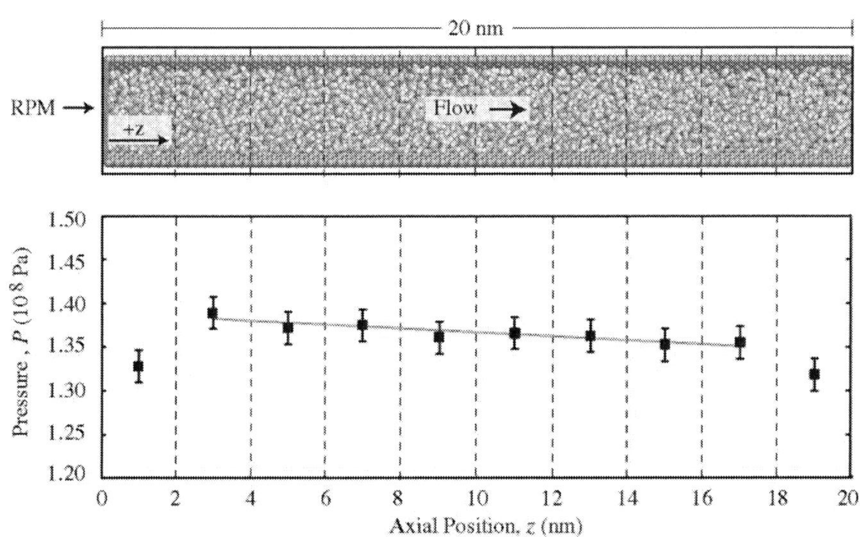

Figure 4.4. Axial pressure gradient inside the 2.77 nm diameter CNT (Thomas John A., McGaughey Alan J. H., 2008).

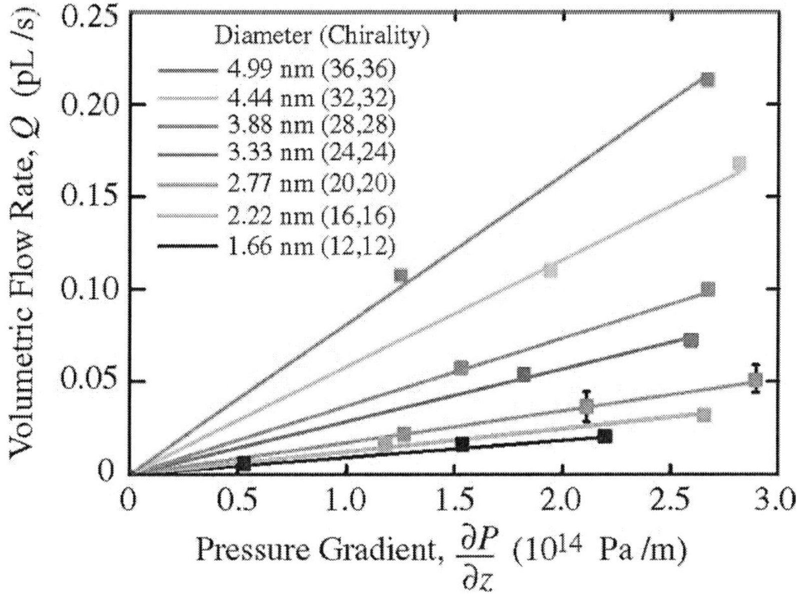

Figure 4.5. Relationship between volumetric liquid flow rate in carbon nanotubes with different diameters and applied pressure gradient (Thomas John A., McGaughey Alan J. H., 2008).

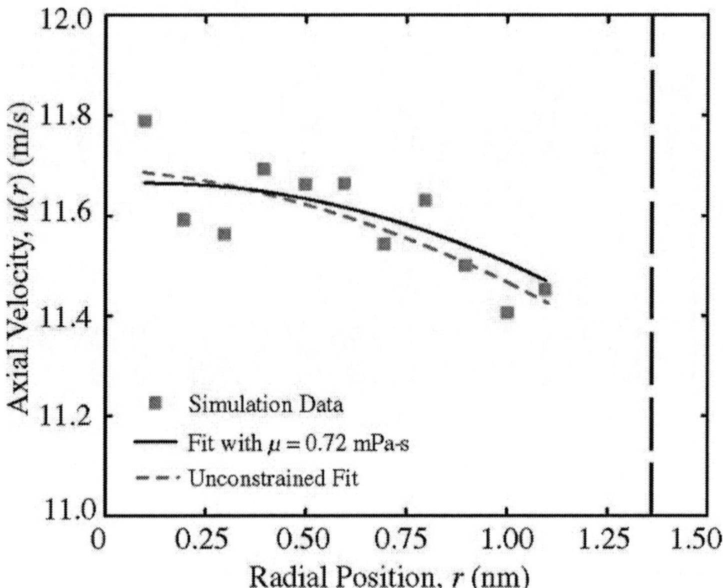

Figure 4.6. Radial velocity profile inside 2.77 nm diameter CNT (Thomas John A., McGaughey Alan J. H., 2008).

In contrast to previous work (Thomas John A., McGaughey Alan JH, 2009) the flow of water under a pressure gradient considered in the single-walled nanotubes of "chair" type of smaller radii: 0.83, 0.97, 1.10, 1.25, 1.39 and 1.66 nm.

Figure 4.7 shows the dependence of the mean flow velocity $\bar{v}$ on the applied pressure gradient $\Delta P / L$ in the long nanotubes is equal to 75 nm at 298 K. A similar picture pattern occurs in the tube with the length of 150 nm.

As one can see, there is conformance with the Darcy law, the average flow rate for each CNT increases with increasing pressure gradient. For a fixed value of $\Delta p / L$, however, the average flow rate does not increase monotonically with increasing diameter of the CNTs, as follows from Poiseuille eqution. Instead, when at the same pressure gradient, decrease of the average speed in a CNT with the radius of 0.83 nm to a CNT with the radius of 1.10 nm, similar to the CNTs 1.10 and 1.25 nm, then increases from a CNT with the radius of 1.25 nm to a CNT of 1.66 nm.

The nonlinearity between $\bar{v}$ and $\Delta P / L$ are the result of inertia losses (i.e. insignificant losses) in the two boundaries of the CNT. Inertial losses depend on the speed and are caused by a sudden expansion, abbreviations, and other obstructions in the flow.

Molecular modeling hn the work of (John A. Thomas, Alan JH McGaughey, Ottoleo Kuter-Arnebeck, 2010), shows that the equation (4.1) (Poiseuille parabola) cdescribes the velocity profile of liquid in a nanotube when the diameter of a flow is 5-10 times more than the diameter of the molecule (≈ 0.17 nm for water).

In figure 4.8, (John A. Thomas, Alan JH McGaughey, 2010), the effect of slip on the velocity profile at the boundary of radius R of the pipe and fluid is shown. When $L_S = 0$ the fluid velocity at the wall vanishes, and the maximum speed (on the tube axis) exceeds flow speed twice.

The figure shows the velocity profiles for Poiseuille flow without slip ($L_S = 0$) and with slippage $L_S = 2R$. The flow rate is normalized to the speed corresponding to the flow without slip. Thick vertical lines indicate the location of the pipe wall. The thick vertical lines indicate the location of the tube wall.

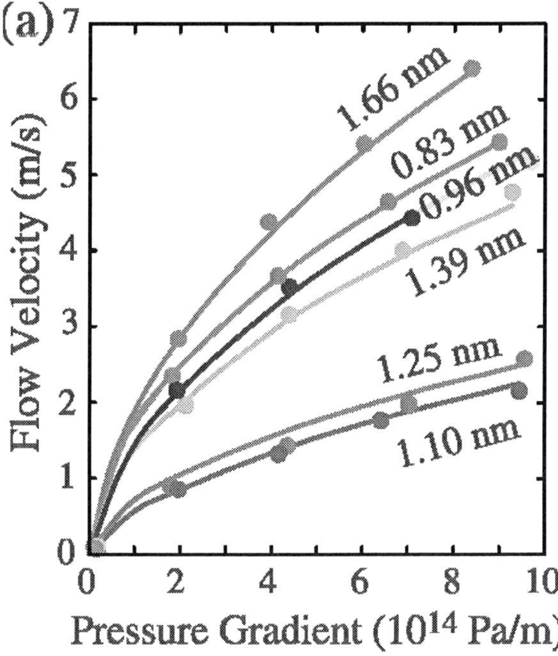

Figure 4.7. Relationship between average flow velocity and applied pressure gradient for the 75 nm long CNTs (Thomas John A., McGaughey Alan J. H., 2009).

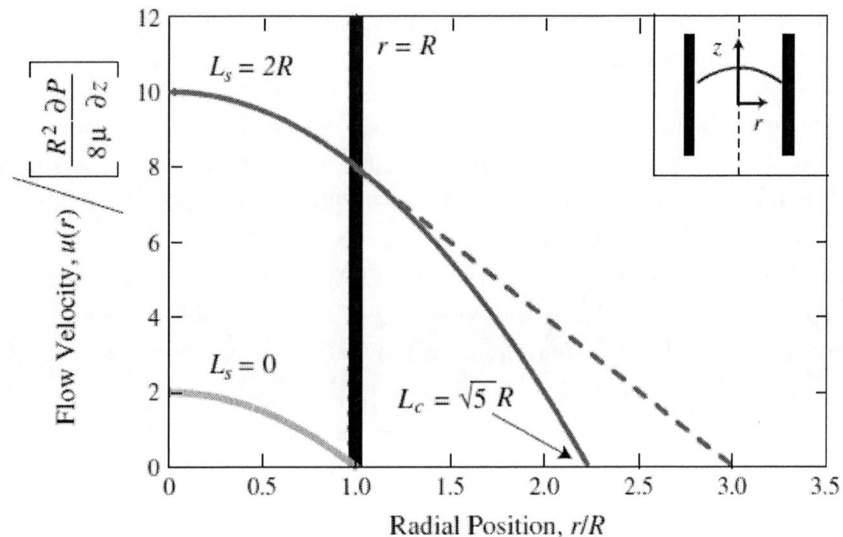

Figure 4.8. No-slip Poiseuille flow and slip Poiseuille flow through a tube.

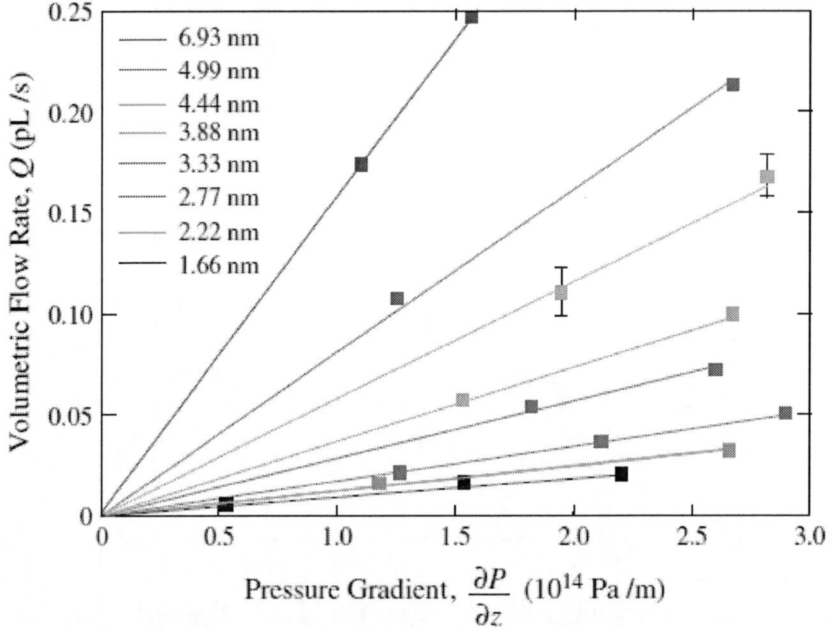

Figure 4.9. Volumetric flow rate in CNTs versus pressure gradient.

Velocity of the liquid on a solid surface can also be quantified by the coefficient of slip L_c. The coefficient of slippage is a difference between the radial position in which the velocity profile would be zero. Slip coefficient is equal to $L_C = \sqrt{R^2 + 2RL_S} = \sqrt{5}R$.

For linear velocity profiles (eg, Couette flow), the length of the slip and slip rate are equal. These values are different for the Poiseuille flow.

Figure 4.9 shows dependence of the volumetric flow rate Q from the pressure gradient $\partial p / \partial z$ in long nanotubes with diameters between 1.66 nm and 6.93 nm. Pressure gradient Q is proportional to $\partial p / \partial z$. As in the Poiseuille flow, volumetric flow rate increases monotonically with the diameter of CNT at a fixed pressure gradient. Magnitudes of calculations error for all the dependencies are similar to the error for the CNT diameter 4.44 nm (marked in the figure).

Researchers (M. Whitby and N. Quirke, 2007) considered steady flow of incompressible fluids in a channel width $2h$ under action of the force of gravity ρg or pressure gradient $\partial p / \partial y$ (which is described by the Navier-Stokes equations). The velocity profile has a parabolic form:

$$U_y(z) = \frac{\rho g}{2\eta} \cdot \left[(\delta + h)^2 - z^2\right]$$

where

δ ; length of the slip, which is equal to the distance from the wall to the point at which the velocity extrapolates to zero.

4.2. MODELING IN NANOHYDROMECANICS

We take into consideration that the mean free path of gas under normal conditions is 65 *нм*, and the distance between the particles - $3,3$ *нм*:

$$\frac{4}{3}\pi\delta^3 n = 1, \qquad \delta_{станд} = 3,3 \text{ нм}, \qquad \lambda = \frac{1}{\sqrt{2}\pi d^2 n}, \qquad \lambda_{станд} = 65 \text{ нм},$$

where;

n ; the concentration of molecules in the air,

Knudsen number $= \lambda/L$,

L ; the characteristic size,

d ; diameter microtubes.

Let's consider the fluid flow through the nanotube. Molecules of a substance in a liquid state are very close to each other (Figure 4.10).

Most liquid's molecules have a diameter of about 0,1 nm. Each molecule of the fluid is "squeezed" on all sides by neighboring molecules and for a period of time $\left(10^{-10} - 10^{-13} s\right)$ fluctuates around certain equilibrium position, which itself from time to time is shifted in distance commensuration with the size of molecules or the average distance between molecules l_{cp}:

$$l_{cp} \approx \sqrt[3]{\frac{1}{n_0}} = \sqrt[3]{\frac{\mu}{N_A \rho}},$$

where;

n_0 ; Number of molecules per unit volume of fluid,

N_A ; Avogadro's number,

ρ ; fluid density

μ ; molar mass

Estimates show that one cubic of nano water contains about 50 molecules. This gives a basis to describe the mass transfer of liquid in a nanotube-based continuum model. However, the specifics of the complexes, consisting of a finite number of molecules, should be kept in mind. These complexes, called clusters in literatures, are intermediately located between the bulk matter and individual particles(atoms or molecules). The fact of heterogeneity of water is now experimentally established (Huang C. et. Al., 2009).

There are groups of molecules in liquid as "micro-crystals" containing tens or hundreds of molecules. Each microcrystal maintains solid form. These groups of molecules or "clusters" exist for a short period of time, then break up and are re-created again. Besides, they are constantly moving so that each molecule does not belong at all times to the same group of molecules, or "cluster".

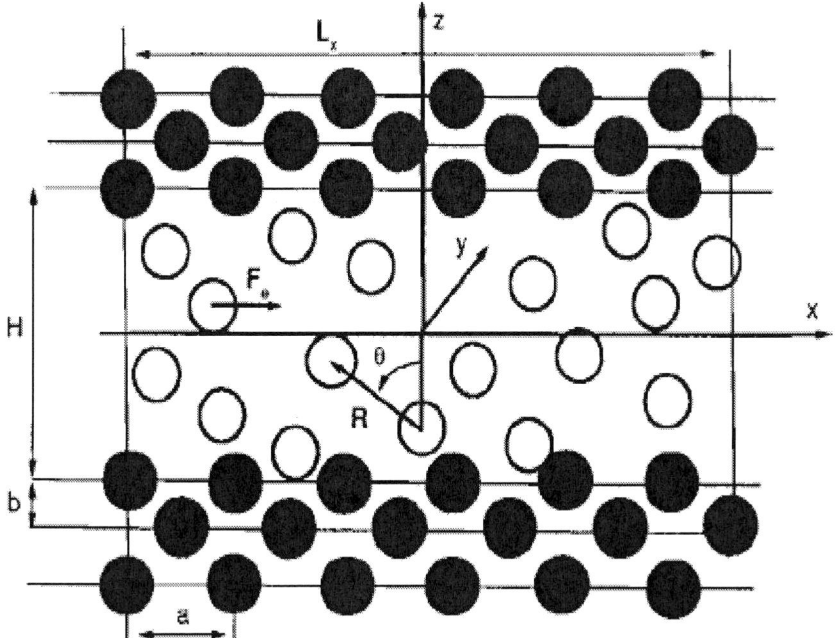

Figure 4.10. The fluid flow through the nanotube.

Modeling predicts that gas molecules bounce off the perfectly smooth inner walls of the nanotubes as billiard balls, and water molecules slide over them without stopping. Possible cause of unusually rapid flow of water is maybe due to the small-diameter nanotube molecules move on them orderly, rarely colliding with each other. This "organized" move is much faster than usual chaotic flow. However, while the mechanism of flow of water and gas through the nanotubes is not very clear and only further experiments and calculations can help understand it.

The model of mass transfer of liquid in a nanotube proposed in this research is based on the availability of nanoscale crystalline clusters in it (Frencel Ya.I., 1941).

A similar concept was developed in (Popov I.Yu. et. al., 2009), in which the model of structured flow of fluid through the nanotube is considered. It is shown that the flow character in the nanotube depends on the relation between the equilibrium crystallite size and the diameter of the nanotube.

Figure 4.11 shows the results of calculations by the molecular dynamics of fluid flow in the nanotube in a plane (a) and three-dimensional state (b). The figure shows the ordered regions of the liquid.

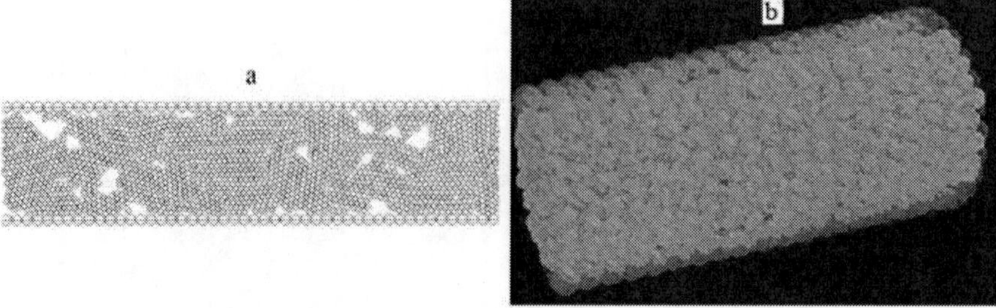

Figure 4.11. The results of calculations of fluid flow in the nanotube (Popov I.Yu. et. al., 2009).

The typical size of crystallite is 1-2 nm, i.e. compared, for example, with a diameter of silica nanotubes of different composition and structure (Popov I.Yu. et. al., 2009).

The flow model proposed in the present work is based on the presence of "quasi-solid" phase in the central part of the nanotube and liquid layer, non-autonomous phases (Gusarov VV and Popov I.Yu., 1996).

Consideration of such a structure that is formed when fluid flows through the nanotube, is also justified by the aforementioned results of the experimental studies and molecular modeling.

When considering the fluid flow with such structure through the nanotube, we will take into account the aspect ratio of "quasi-solid" phase and the diameter of the nanotube. Then the character of the flow is stable and the liquid phase can be regarded as a continuous medium with viscosity η.

Let's establish relationship between the volumetric flow rate of liquid Q flowing from a liquid layer of the nanotube length l, the radius R and the pressure drop $\Delta p / l$, $\Delta p = p - p_0$, where;

p_0 is the initial pressure in the tube (Figure 4.12).

Let R_0 be a radius of the tube from the "quasi-solid" phase and;

v ; velocity of fluid flow through the nanotube.

Structural regime of fluid flow (figure 4.13) implies the existence of the continuous laminar layer of liquid (the liquid layer in the nanotube) along the walls of a pipe. In the central part of a pipe a core of the flow is observed, where the fluid moves, keeping his former structure, i.e. as a solid ("quasi-solid" phase in the nanotube). The velocity slip is indicated in Figure 4.13a through v_0.

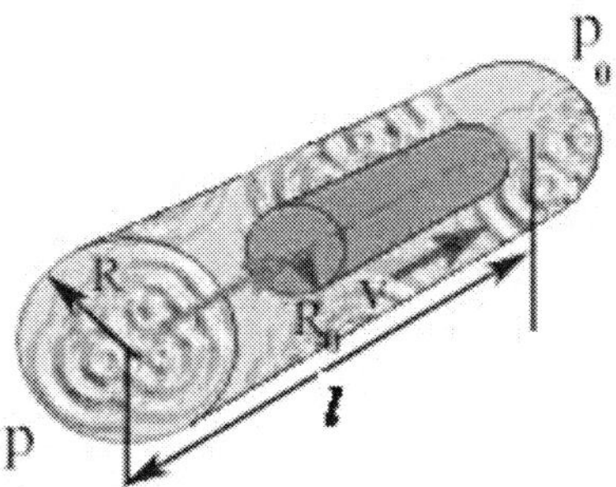

Figure 4.12. Flow through liquid layer of the nanotube.

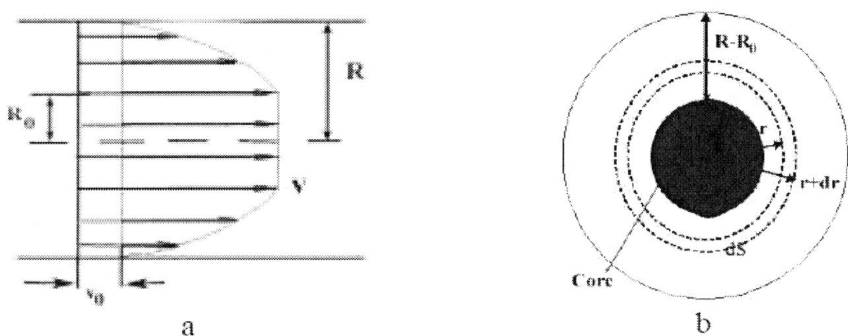

Figure 4.13. Structure of the flow in the nanotube.

Let's find the velocity profile $v(r)$ in a liquid interlayer $R_0 \le r \le R$ of the nanotube. We select a cylinder with radius r and length l in the interlayer, located symmetrically to the center line of the pipe (see figure 4.13б).

At the steady flow, the sum of all forces acting on all the volumes of fluid with effective viscosity η, is zero.

The following forces are applied on the chosen cylinder: the pressure force and viscous friction force affects the side of the cylinder with radius r, calculated by the Newton formula.

Thus,

$$(p - p_0)\pi r^2 = -\eta \frac{dv}{dr} 2\pi r l \tag{4.6}$$

Integrating expression (4.6) between r to R with the boundary conditions $r = R : v = v_0$, we obtained a formula to calculate the velocity of the liquid layers located at a distance r from the axis of the tube:

$$v(r) = (p - p_0)\frac{R^2 - r^2}{4\eta l} + v_0 \tag{4.7}$$

Maximum speed $v_я$ has the core of the nanotube $0 \le r \le R_0$ and is equal to:

$$v_я = (p - p_0)\frac{R^2 - R_0^2}{4\eta l} + v_0 \tag{4.8}$$

Such structure of the liquid flow through nanotubes considering the slip is similar to a behavior of viscoplastic liquids in the tubes. For viscoplastic fluids a characteristic feature is that they are to achieve a certain critical internal shear stresses τ_0 and behave like solids. Meanwhile, when internal stress exceeds a critical value begin to move as normal fluid. In the work of (Popov I.Yu. et. Al., 2009) the liquid behaves in the nanotube the similar way. A critical pressure drop is also needed to start the flow of liquid in a nanotube.

Structural regime of fluid flow requires existance of continuous laminar layer of liquid along the walls of pipe. In the central part of the pipe is observed flow with core radius R_{00}, where the fluid moves, keeping his former structure (i.e. as a solid).

The velocity distribution over the pipe section with radius R of laminar layer of viscoplastic fluid is expressed as follows:

$$v(r) = \frac{\Delta p}{4\eta l}\left(R^2 - r^2\right) - \frac{\tau_0}{\eta}(R - r) \tag{4.9}$$

The speed of flow core in $0 \le r \le R_{00}$ is equal

$$v_я = \frac{\Delta p}{4\eta l}\left(R^2 - R_{00}^2\right) - \frac{\tau_0}{\eta}\left(R - R_{00}\right) \tag{4.10}$$

Let's calculate the flow or quantity of fluid flowing through the nanotube cross-section S at a time unit. The liquid flow dQ for the inhomogeneous velocity field flowing from the cylindrical layer of thickness dr, which is located at a distance r from the tube axis is determined from the relation

$$dQ = v(r)dS = v(r)2\pi r dr \tag{4.11}$$

where dS - the area of the cross-section of cylindrical layer (between the dotted lines in figure 4.13).

Let's place equation (4.7) in (4.11), integrate over the radius of all sections from R_0 to R and take into account that the fluid flow through the core flow is determined from the relationship $Q_я = \pi R_0^2 v_я$. Then we get the formula for the flow of liquid from the nanotube:

$$Q = \pi R^2 v_0 + Q_P\left[1 - \left(\frac{R_0}{R}\right)^4\right] \tag{4.12}$$

If $(R_0 / R)^4 << 1$ (no nucleus) and $v_0 \Delta p R^2 / 8l\eta << 1$ (no slip), then (4.12) coincides with Poiseuille formula (3.5). When $R_0 \approx R$ (no of a viscous liquid interlayer in the nanotube), the flow rate Q is equal to volumetric flow $Q \approx \pi R^2 v_0$ of fluid for a uniform field of velocity (full slip).

Accordingly, flow rate of the viscoplastic fluid flowing with a velocity (4.7), is equal to:

$$Q = -\frac{\pi R^3 \tau_0}{3\eta}\left[1 - \left(\frac{R_{00}}{R}\right)^3\right] + Q_P\left[1 - \left(\frac{R_{00}}{R}\right)^4\right] \tag{4.13}$$

Comparing equations (4.7), (4.8), (4.9), (4.10) and (4.12), (4.13), we can see that the structure of the flow of the liquid through the nanotubes considering the slippage, is similar to that of the flow of viscoplastic fluid in a pipe of the same radius R.

Given that the size of the central core flow of viscoplastic fluid (radius R_{00}) is defined by

$$R_{00} = \frac{2\tau_0 l}{\Delta p} \tag{4.14}$$

for viscoplastic fluid flow we obtain Buckingham formula:

$$Q = Q_P\left[1 + \frac{1}{3}\left(\frac{2l\tau_0}{R\Delta p}\right)^4 - \frac{4}{3}\left(\frac{2l\tau_0}{R\Delta p}\right)\right] \tag{4.15}$$

We'll Establish a conformity of the pipe, that implements the flow of a viscoplastic fluid with a fluid-filled nanotube, the same size and with the same pressure drop. We say that an effective internal critical shear stress τ_{0ef} of viscoplastic fluid flow, which ensures the coincidence rate with the flow of fluid in the nanotube. Then from (4.15) we obtain equation of fourth order to determine τ_{0ef}:

$$\left(\frac{2l\tau_{0ef}}{R\Delta p}\right)^4 - 4\left(\frac{2l\tau_{0ef}}{R\Delta p}\right) = A, \quad A = 3(\varepsilon - 1), \quad \varepsilon = Q/Q_P, \tag{4.16}$$

The solution of equation (4.16) can be found, for example, the iteration method of Newton:

$$\bar{\tau}_{0ef\,n} = \bar{\tau}_{0ef\,n-1} - \frac{\bar{\tau}^{4}_{0ef\,n-1} - 4\bar{\tau}_{0ef\,n-1} - A}{4\bar{\tau}^{3}_{0ef\,n-1} - 4}, \quad \bar{\tau}_{0ef} = \frac{2l\tau_{0ef}}{R\Delta p}, \tag{4.17}$$

The first component in (4.12) represents the contribution to the fluid flow due to the slippage, and it becomes clear that the slippage significantly enhances the flow rate in the nanotube, when $l\eta v_0 >\approx \Delta pR^2$.

This result is consistent with experimental and theoretical results of (Kalra A. et al. 2003), (Majumder M. et al., 2005), (Skoulidas AI et al. 2002), (Hummer G. et al. 2001), (Holt JK et al. 2006), which show that water flow in nanochannels can be much higher than under the same conditions, but for the liquid continuum.

In the absence of slippage $\varepsilon = 1$ the equation (4.16) has a trivial solution $\bar{\tau}_{0ef} = 0$.

The Results of the Calculations

Let`s determine the dependence of the effective critical inner shear stress τ_{0ef} on the radius of the nanotubes, by taking necessary values for calculations $\varepsilon = Q/Q_P$ from the work of (Thomas John A. and McGaughey Alan JH, 2008). The results of calculations at $\Delta p/l = 2{,}1 \cdot 10^{14}\ Pa/m$ are in the table below:

R, м	τ_{0ef} (Па)	$\varepsilon = Q/Q_P$
$0{,}83 \cdot 10^{-9}$	498498	350
$1{,}11 \cdot 10^{-9}$	577500	200
$1{,}385 \cdot 10^{-9}$	632599	114
$1{,}665 \cdot 10^{-9}$	699300	84
$1{,}94 \cdot 10^{-9}$	782208	68
$2{,}22 \cdot 10^{-9}$	855477	57
$2{,}495 \cdot 10^{-9}$	932631	50

Calculations show that the value of effective internal shear stress depends on the size of the nanotube.

Figure 4.14 shows the dependence τ_{0ef} on the nanotube radius.

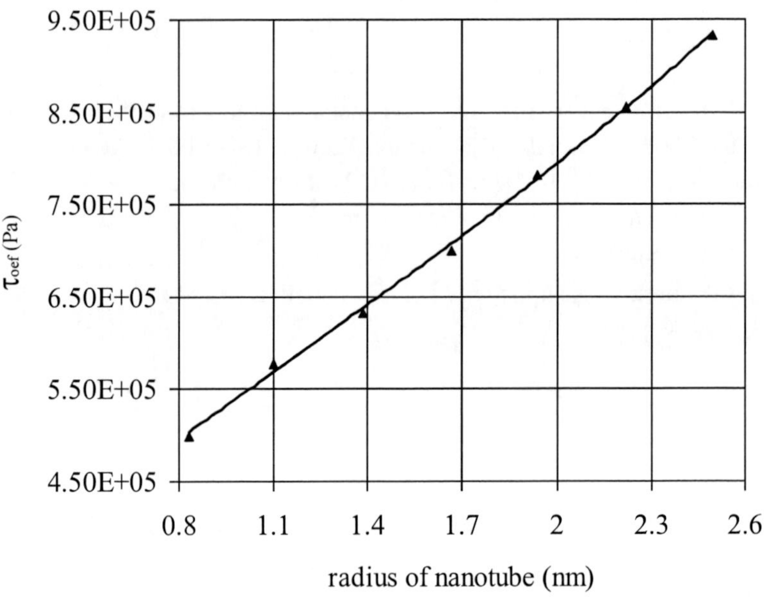

Figure 4.14. Dependence of the effective inner shear stress from the radius of the nanotube.

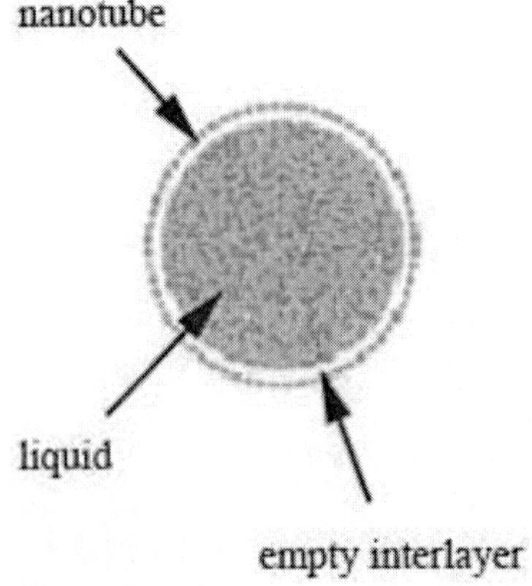

Figure 4.15. The structure of the flow.

The dependence $\tau_{0ef}(R)$ is almost linear. Within the range of the considered nanotube sizes τ_{0ef} has a relatively low value, which indicates the smoothness of the surface of carbon nanotubes.

The Flow of Fluid with an Empty Interlayer

The works of (EM Kotsalis et al., 2004) and (Xi Chen et al. 2008) were analyzed in the aforementioned analysis of the structure of liquid flow in carbon nanotubes. The results of the calculations of the cited works (figure 4.6 and 4.7) showed that during the flow of the liquid particles, an empty layer between the fluid and the nanotube is formed. The area near the walls of the carbon nanotube $R_* \leq r \leq R$ becomes inaccessible for the molecules of the liquid due to van der Waals repulsion forces of the heterogeneous particles of the carbon and water (figure 2). Moreover, according to the results of (EM Kotsalis et al., 2004) and (Xi Chen et al. 2008) thicknesses of the layers $R_* \leq r \leq R$ regardless of radiuses of the nanotubes are practically identical: $R_* / R \approx 0{,}88$.

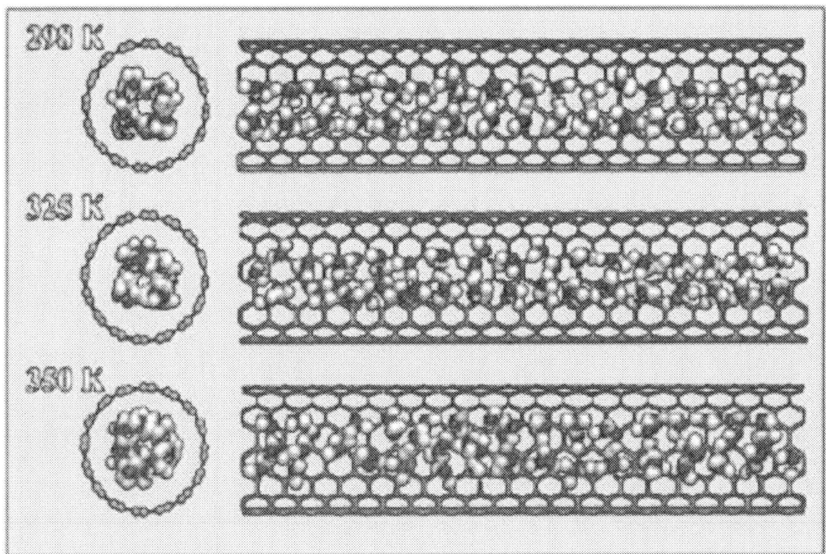

Figure 4.16. The configuration of water molecules inside single-walled carbon nanotubes.

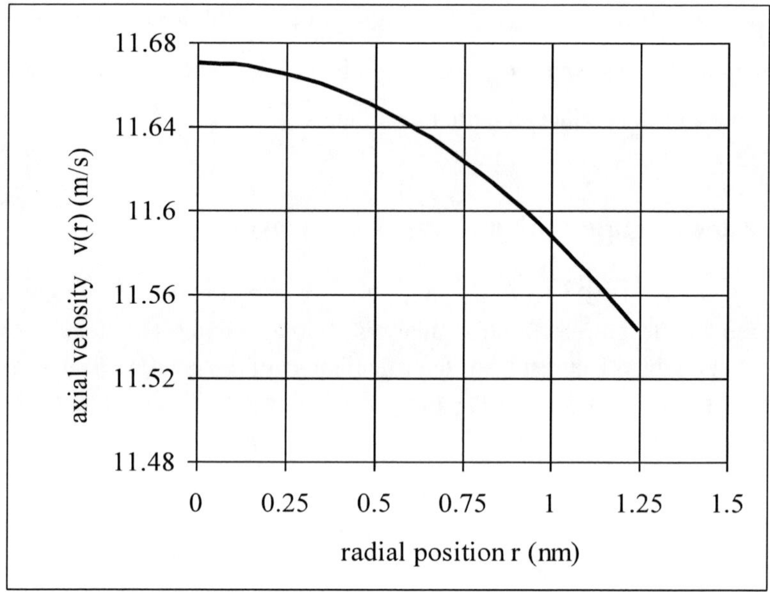

Figure 4.16. Profile of the radial velocity of water in carbon nanotubes.

A similar result was obtained in (Hongfei Ye et al, 2011), which is an image (figure 4.16) of the configuration of water molecules inside (8, 8) single-walled carbon nanotubes at different temperatures: 298, 325, and 350 ^0K.

Integrating expression (4.6) between r to R_* at the boundary conditions $r = R_* : v = v_*$, we obtain a formula to calculate the velocity of the liquid layers located at a distance r from the axis of the tube:

$$v(r) = v_* + \frac{2Q_p}{\pi R^2}\left[\left(\frac{R_*}{R}\right)^2 - \left(\frac{r}{R}\right)^2\right] \tag{4.18}$$

Let's insert (4.18) in (4.11), integrate over the radius of all sections from 0 to R_*. Then we get a formula for the flow of liquid from the nanotube:

$$Q = Q_P \left[v_* \frac{8l\eta R_*^2}{\Delta p R^4} + \left(\frac{R_*}{R} \right)^4 \right] \tag{4.19}$$

or

$$\varepsilon = \frac{8l\eta v_*}{\Delta p R^2} \left(\frac{R_*}{R} \right)^2 + \left(\frac{R_*}{R} \right)^4$$

from which we can determine the unknown v_*:

$$v_* = \frac{Q_P}{\pi R^2} \left[\varepsilon - \left(\frac{R_*}{R} \right)^4 \right] \left(\frac{R}{R_*} \right)^2 \tag{4.20}$$

Figure 4.16 shows the profile of the radial velocity of water particles in a carbon nanotube with a diameter of 2.77 nm, calculated using the formula (4.18) at $Q_P = 4{,}75 \cdot 10^{-19} n/m^2$, $\varepsilon = 114$. The velocity at the border v_* is equal to 11,55 m/s. It is seen that the velocity profile is similar to a parabolic shape, and at the same time agrees with the calculations of (Thomas John A. and McGaughey Alan JH, 2008), obtained by using the model of molecular dynamics.

The calculations suggest the following conclusions. Flow of liquid in a nanotube was investigated using synthesis of the methods of the continuum theory and molecular dynamics. Two models are considered. The first is based on the fact that fluid in the nanotube behaves like a viscoplastic. A method of calculating the value of limiting shear stress is proposed, which was dependent on the nanotube radius. A simplified model agrees quite well with the results of the molecular simulations of fluid flow in carbon nanotubes. The second model assumes the existence of an empty interlayer between the liquid molecules and wall of the nanotube. This formulation of the task is based on the results of experimental works known from the literature. The velocity profile of fluid flowing in the nanotube is practically identical to the profile determined by molecular modeling.

As seen from the results of the calculations, the velocity value varies slightly along the radius of the nanotube. Such a velocity distribution of the fluid particles can be explained by the lack of friction between the molecules of the liquid and the wall due to the presence of an empty layer. This leads to an easy slippage of the liquid and, consequently, anomalous increase in flow compared to the Poiseuille flow.

Chapter 5

NANOHYDROMECHANICS

5.1. NANOPHENOMENON IN OIL PRODUCTION

Oil-saturated layers are porous materials with different pore sizes, pore channels and composition of rocks that define the features of interaction formation and injected fluids with the rock. Taking the mentioned into account we can conclude that the displacement of oil from oil fields in production wells is not a mechanical process of substitution of oil displacing it with water, but a complex physical-chemical process in which the decisive role is played by the phenomenon of ion exchange between reservoir and injected fluids with the rock, i.e. nanoscale phenomena.

The mechanism of displacement of oil in the reservoir and its recovery is largely determined by the molecular-surface processes occurring at phase interfaces (the rock-forming minerals - saturate the reservoir fluids and gases - displacing agents). Therefore the problem of wetability is one of the major problems in oil and gas field of nanoscience.

As the clay is an ultra-system, a huge amount of research on regulation of the position of clay minerals in porous media with good reason can be attributed to nanoscience. To it also should be included the study of gas hydrates, a number of processes regulating the properties of the pumped oil and gas trapped water, water - oil preparation.

Filtering oil in reservoirs at a depth of 13 km of hard rock is determined by the hydrodynamics of the Darcy law. Of the Navier-Stokes equation

$$\rho\left(\frac{v}{t}+(\nabla\cdot v)v\right)=-\nabla p+\mu v, \tag{5.1}$$

At very low Reynolds numbers it follows that we can neglect the inertial forces and simplify viscous friction:

$$-\nabla p-\mu\frac{v}{d^2}=0, \tag{5.2}$$

d — characteristic pore size. Equation (5.2) yields Darcy's law

$$v=-\frac{K}{\mu}\nabla p, \tag{5.3}$$

where

$K=d^2$ — permeability of the medium, a μ — viscosity of the fluid.

Typical permeability values range from $5\ mD$ to $500\,mD$. The permeability of coarse-grained sandstone is $10^{-8}-10^{-9}\,sm^2$, the permeability of dense sandstone around $10^{-2}\,sm^2$. Medium with the permeability $1\,D$ passes of fluid flow with a viscosity 1 sP at a pressure gradient $1aym/\,sm$.

During filtration oil fills and moves in the pores with the size of $1mkm$. In order to displace oil from the reservoir one should have a medium with density and viscosity of oil and the size of $1mkm$.

It seems that it is necessary to conduct experimental and theoretical studies of possible ways to obtain microbubbles nanoscale environments, to make the calculations and estimates of energy costs upon receipt of such media in different ways and their application to problems of oil production.

5.2. Petroleum Composition

Chemically, oil is a complex mixture of hydrocarbons (HC) and carbon compounds. It consists of the following elements: carbon (84-87%), hydrogen

(12-14%), oxygen, nitrogen, sulfur (1-2%). The sulfur content can reach up to 3-5%. Oils can contain the following parts: a hydrocarbon, asvalto-resinous, porphyrins, sulfur and ash. Oil has a dissolved gas that is released when it comes to the earth's surface.

The main part of petroleum hydrocarbons are different in their composition, structure and properties, which may be in gaseous, liquid and solid state. Depending on the structure of the molecules they are classified into three classes - paraffinic, naphthenic and aromatic. But a considerable proportion of oil is hydrocarbons of mixed structure containing structural elements of all three above-mentioned classes. The structure of the molecules determines their chemical and physical properties.

Carbon is characterized by its ability to form chains in which the atoms are connected in series with each other. In remaining connections hydrogen atoms are attached to the carbon. The number of carbon atoms in the molecules of paraffinic hydrocarbons exceeds the number of hydrogen atoms twice, with some constant excess in all the molecules equal to 2. In other words, the general formula of this class of hydrocarbons is C_nH_{2n+2}. Paraffinic hydrocarbons chemically more stable and refer to the limiting HC.

Depending on the number of carbon atoms in the molecule hydrocarbons may be in one of the three states of aggregation. For example, if there are one to four carbon atoms in a molecule (CH_4-C_4H_{10}), the hydrocarbon is a gas, from 5 to 16 (C_5H_{16} – $C_{16}H_{34}$) - a liquid hydrocarbon, and if more than 16 ($C_{17}H_{36}$ и т.д.) - solid.

Thus, paraffin hydrocarbons in oil can be represented by gases, liquids and solid crystalline substances. They have different effects on the properties of oil: gas reduces viscosity and increases the vapor pressure.

Fluid paraffins dissolve well in oil only at elevated temperatures, forming a homogeneous mixture. Hard paraffins also dissolve well in oil forming the true molecular mixtures. Paraffin hydrocarbons (with the exception of ceresin) can be easily crystallized in the form of plates and plate strips.

Naphthenic (tsiklanovae or alicyclic) hydrocarbons have cyclic structure (C/C_nH_{2n}), to be exact, they are composed of several groups - CH_2 - interconnected in ringed system. Oil contains mainly naphthenes consisting of five or six groups of CH_2. All connections of carbon and hydrogen are saturated, so the naphthenic oil has stable properties. Compared with paraffin, naphthenes have a higher density and lower vapor pressure and have better solvent power.

Aromatic hydrocarbons (arena) are represented by the formula C_nH_n, are most poor by hydrogen. The molecule has a form of a ring with unsaturated carbon connections. The simplest representative of this class of hydrocarbons is benzene C_6H_6, which consists of six groups of CH. For aromatic hydrocarbons a large solubility, higher density and boiling point are typical.

Asphalt-resinous portion of oil is a substance of dark color, which is partially soluble in gasoline. They have the ability to swell in solvents, and then pass into mixture.

The solubility of asphaltenes in the resin-carbon systems increases with decreasing concentration of light hydrocarbons and increasing concentrations of aromatic hydrocarbons.

The resin does not dissolve in gasoline and is a polar substance with a relative molecular mass of 500-1200. They contain the bulk of oxygen, sulfur and nitrogen compounds of oil.

Asphaltic-resinous substances, and other polar components are surface-active compounds and natural oil-water emulsion stabilizers.

Special nitrogenous compounds of organic origin are called porphyrins. It is assumed that they were formed from animal hemoglobin and chlorophyll of plants. These compounds are destroyed at temperatures of 200-250 ^0C .

Sulfur is prevalent in petroleum and hydrocarbon gas and is contained both in the eree State and in the form of compounds (hydrogen sulphide, mercaptans).

Ash is the residue which is formed by burning oil. This is a different mineral compound, usually iron, nickel, vanadium, and sometimes sodium.

Properties of oil determine the direction reprocessing and affect the products derived from petroleum, so there are different types of classification, which reflect the chemical nature of oil and determine possible areas of processing.

For example, in the base of the classification, reflecting the chemical composition is laid the preference content of one or more classes of hydrocarbons in the oil.

Naphthene, paraffin, paraffin-naphthene, paraffin-naphthene-aromatic, naphthene-aromatic and aromatic hydrocarbons are being distinguished. Thus, all fractions in the paraffin oils contain a significant quantity of alkanes.

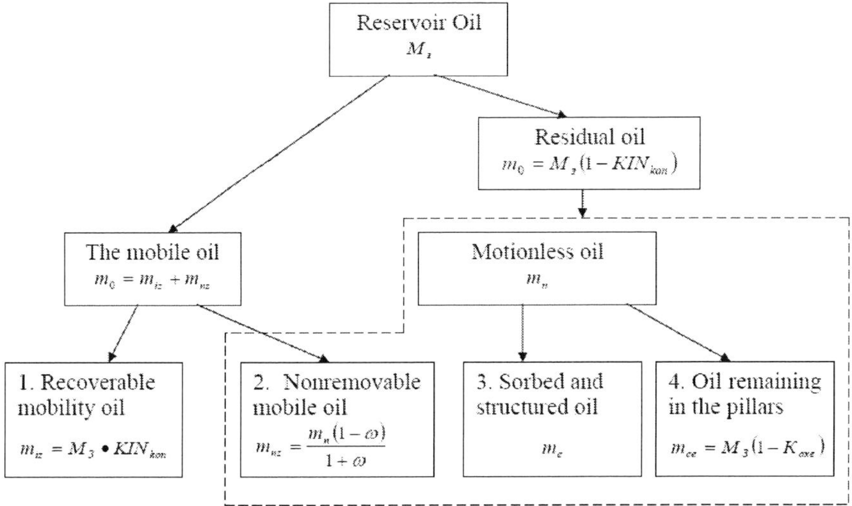

Figure 5.1. The constituents of reservoir oil.

In the paraffin-naphthene-aromatic hydrocarbons of all three classes are contained in approximately equal amounts. Naphthene-aromatic oil is characterized mainly by the content of cycloalkanes and arenes, especially in the heavy fractions.

Classification is also used by the content of asphaltenes and resins.

In the technical classification oils are divided into classes according to the sulfur content;

Types; by the output of factions at certain temperatures

Groups; by the potential content of base oils

Species; by content of solid alkanes (of papafins)

Figure 5.1 shows the components of the reservoir oil, which have different average integrated over the period of development and the entire volume of the reservoir values of physico-chemical properties. Here;

$M_з$; reserves of reservoir oil at reservoir conditions,

m_B ; Mass of water in the reservoir area drained by the end of development, t;

ω ; watering at the end of development, the proportion of units;

$m_П$; Mass of mobile oil, t;

$m_{из}$; Mass of extracted oil, t;

$m_{нз}$; Mass of the nonremovable movable oil, nonremovable t;

m_c ; Mass of the adsorbed and structured oil t;

$m_{Ц}$; Mass of oil remaining in pillars, t.

The mobile oil ($m_{П}$) is part of the reservoir oil moving along layer due to the impact of external influences.

Recoverable mobile oil ($m_{из}$) is certain part of the mobile oil, which can be extracted from the reservoir as a result of industrial activity with taking into account of economic and technological limitations.

Nonremovable mobile oil ($m_{нз}$) is part of the mobile oil, which will not be extracted from the reservoir using the technologies as a result of industrial activity on the economic and technological constraints.

The residual oil (m_o) is part of the reservoir oil, located in the reservoir at the end of the displacement.

Motionless oil ($m_н$) is part of the reservoir oil, remaining motionless in the reservoir due to external influence.

Sorbed and structured oil (m_c) is part of motionless oil, retained near the surface of the collector by the intermolecular interaction.

Oil remaining in pillars ($m_{Ц}$) is part of motionless oil, is not involved in the process of drainage.

Based on the proposed separation of produced oil into separate components, the average integral value of the physico-chemical properties of produced oil $\overline{X_з}$ for counting of reserves must be found from the expression:

$$\overline{X_з} = \frac{X_{из} \cdot m_{из} + X_{нз} \cdot m_{нз} + X_c \cdot m_c + X_{Ц} \cdot m_{Ц}}{M_з} \quad ; \qquad (5.4)$$

$$M_з = m_{из} + m_{нз} + m_c + m_{Ц} \qquad (5.5)$$

or

$$X_з = X_{из} d_{из} + X_{нз} d_{нз} + X_c d_c + X_{Ц} d_{Ц} \qquad (5.6)$$

where;

$X_{из}$, $X_{нз}$, X_c, $X_Ц$; the mean of the integral value of the corresponding component of reservoir oil;

$d_{из}$, $d_{нз}$, d_c, $d_Ц$; the mass fraction of the corresponding component of reservoir oil $d_i = m_i / M_з$,

where;

m_i; The mass of i component of reservoir oil,

$M_з$; The mass of reservoir oil (geological reserves of oil reservoir).

Formulas (5.4) and (5.6) allow us to calculate the average integral value of the physico-chemical properties of reservoir oil by using a different source (relative or absolute).

The average integral value of the physico-chemical properties of mobile oil X_n for use in the calculation of oil displacement processes must be found out of the expression:

$$\overline{X_n} = \frac{X_{из} \cdot m_{из}}{m_n} + \frac{X_{нз} \cdot m_{нз}}{m_n}; \tag{5.7}$$

or

$$\overline{X_n} = X_{из} c_{из} + X_{нз} c_{нз} \tag{5.8}$$

where;

$X_{из} c_{из} = m_{нз} / m_n$; property value and the mass fraction of component of recovered part of mobile oil, respectively;

$X_{нз} c_{нз} = m_{нз} / m_n$; Property value and the mass fraction of component of nonremovable part of mobile oil, respectively.

In practice, there are a number of contradictions in the calculation methods of geological reserves of hydrocarbons and the calculation of oil recovery processes. For example, an anachronism in the method of counting, which is based on the volume (sealer) ratio of produced oil, because the value of reserves is obtained depending on the conditions of oil, since the magnitude of the volume ratio is dependent on the parameters of technology of oil preparation (the number of stages of oil separation and the temperature and pressure conditions).

The density is determined for each formation zone in the case of inhomogeneity of the properties of reservoir oil in the layer. In this case, during the separation of the extracted products on the commodity oil and associated gas only their masses will be dependent on the properties of reservoir oil and the parameters of ground preparation.

5.3. Nanobubbles of Gas, Oil in the Pore Channels and Water

The average diameter of the channels of the porous medium d is easy to estimate using the well known relationship

$$d = \frac{4m}{S} \tag{5.9}$$

where;

m ; Porosity of the medium, fraction of a unit;

S ; Specific surface (surface per a volume)

Clay Oil Gas Motherboard Thicknesses (OGMT) usually characterized by a porosity of 10-20%, and their specific surface of not less than $10^8 m^{-1}$. Consequently, the diameters of the pore channels of clay OGMT not exceed an average of 3 nm.

The minimum diameter of a gas bubble in a water medium can be calculated from the following assumptions:

The gas pressure p_g in a bubble caused by the action of two components: of the deposit pressure $p_{пл}$ and the pressure of surface tension p_σ .

$$p_g = p_{пл} + p_\sigma \tag{5.10}$$

or in accordance with the law of Laplace

$$p_g = p_{nл} + \frac{2\sigma}{R} \quad (5.11)$$

where;

σ ; Coefficient of the surface tension of the liquid, in which formed a gas bubble; R ; the bubble radius (R.I. Nigmatulin, 1987) (F.B. Nagiyev, 1985).

Gas bubbles can be considered an ensemble of hydrocarbon molecules with a mass equal to or less than the buoyant force acting on it in the liquid, which gives a formal record

$$NMg = \rho Wg \quad (5.12)$$

where;

N ; Number of molecules in the ensemble;

M ; Mass of each molecule;

g ; Acceleration of free fall;

ρ ; Density the liquid;

W ; Volume of the bubble

The molecular energy of the ensemble, keeping a bottle from collapse (i.e., from the dissolution of gas), is defined by the van der Waals forces, in which force of intermolecular interaction can be ignored, since the gas has virtually no internal pressure. In this case, the magnitude of this energy per unit volume of a gas bubble is equal to the pressure p_0 in the bubble:

$$p_0 = \frac{NkT}{W - W_0} \quad (5.13)$$

where;

T ; Absolute temperature of the gas ${}^0K$;

k ; Boltzmann's constant;

$W_0 = Nw_0$;

$w_0 = 12w$; Volume per one molecule of gas at the critical temperature and pressure; w ; the real volume of a gas molecule

From (5.13) after simple transformations, with taking into account expression for w_0 we have

$$N = \frac{p_0 W}{kT + p_0 w_0} \quad (5.14)$$

which after substitution in equation (5.12) gives

$$p_0 = \frac{kT}{M / \rho - w_0} \quad (5.15)$$

The condition of the bubbles floating necessarily implies the existence of the phase boundary, i.e. surface bounding the volume of the bubble. A necessary and sufficient condition for such a boundary is the equality of pressures in gas and liquid phases:

$$p = -p_0 \quad (5.16)$$

Substituting the values of pressures from (5.11) and (5.15) gives the desired diameter of the bubble in the form of

$$d_n = 2R = \frac{4\sigma(Vp_0 - M / \rho)}{kT + p_{пл}(w_0 - M / \rho)} \quad (5.17)$$

Calculation shows that the diameter of the molecule the simplest hydrocarbon gas - methane is equal to 0.38 nm, $w_0 = 0{,}345 \cdot 1^{-27}\, m^3$.

If this take into account, that $M = 27{,}2 \cdot 10^{-27}\, kg, \rho \sim 10^3\, kg / m^3$, $k = 1{,}38 \cdot 10^{-23}\, J/^0K$, $T = 360^0 K$, $\sigma = 62{,}3 \cdot 10^{-3}\, n / m$, then from formula (5.17) is easy to determine the value of the minimum diameter of the gas bubble.

Under hydrostatic pressure of petroleum $p_{пл}$ = 20 Mpa, it is 7 nm. This is more than twice the diameter of the pore channels. Consequently, in the cramped conditions of the pore space of OGMT formation of a gas bubble is impossible, because the capillary pressure in the ducts with a diameter of 3 nm to more than 5 times higher than the pressure p_0 at a given temperature. This violates the necessary condition for the existence of a gas bubble in a liquid (5.16) and prevents phase separation, since the external pressure leads to a collapse of gas bubbles, ie, to its "dissolution" of the pore fluid, if it ocuures at all.

It is clear that the required number of methane molecules to form a bubble is equal to

$$N = \frac{W_n}{w_0} = \frac{179{,}6 \cdot 10^{-27}}{0{,}345 \cdot 10^{-27}} = 520$$

where;

W_n ; Volume of the bubble with diameter of 7 nm

The minimum diameter of a bubble of oil in water is calculated based on approximately similar considerations.

A bubble of oil in an aqueous medium can be considered as an ensemble of molecules, which are able to form an interface between the liquid phases. The above (average) radius of the volume of such an ensemble can be obtained from (5.11):

$$R = \frac{2\sigma}{p_g - p_{nл}} \tag{5.18}$$

where;

σ ; Border tension coefficient in the water- oil, equal to approximately $45 \cdot 10^{-3} n/m$; $p_g = p_{nл} + p$,

where;

p ; additional molecular pressure in the equation of van der Waals forces, which for a liquid at condition of it incompressibility it is possible to express from the of the same equation of van der Waals forces:

$$p = -\left(\frac{kT}{w} + p_{nл}\right) \tag{5.19}$$

where;

w ; The real volume of one molecule of the liquid, and "minus" sign due to the fact that the pressure force directed toward the center of the bubble of oil.

Then from (5.18) and (5.19) we have finally

$$d_n = 2R = \frac{4\sigma w}{kT + wp_{пл}} \tag{5.20}$$

The resulting formula determines the minimum value of the diameter of the bubble of oil in a water medium. In this case the real volume of the hydrocarbon molecule with one carbon atom - methane, $w \sim 3{,}5 \cdot 10^{-28}\, м^3$.

The total mass of hydrocarbons that are brought by failution flow per unit area of contact OGMT - rock reservoir is defined as

$$G = \int_0^t V_G dt \tag{5.21}$$

where;

V_G ; Velocity of hydrocarbon generation Organic Materials (OM) from OGMT;

t - Time;

G ; The productivity of generation

The formation of hydrocarbons from the Dispersed Organic Material (DOM) can be regarded as an elimination process in which the initial component is the reactive part of the DOM, and the final product (hydrocarbon molecules). Then, the reaction velocity of elimination is written well-known formula

$$V = \varepsilon(\Gamma - x) = \varepsilon\Gamma e^{-\varepsilon t} \tag{5.22}$$

since $x = \Gamma\left(1 - e^{-\varepsilon t}\right)$.

In such a formulation;

V ; The velocity of formation hydrocarbon from DOM;

ε ; Integrated constant of reaction velocity,

Γ ; The initial concentration of the reaction capable of part of the DOM in the breed,

x ; the mass of hydrocarbons that has developed in time t.

ε ; is usually determined experimentally from the velocity of formation of hydrocarbon out from the given DOM at different temperatures.

Equation (5.22) is valid for open systems in which the products of the reaction easily draw off the center of the reaction. In conditions of porous medium of natural OGMT with a limited pore volume it is necessary to introduce a factor considering the difficulty of derivatives removal (reaction products) into the formula (5.22).

As noted above, gas (or oil) can not exist in the pore channels of clay medium in the form of separate phase, the elimination reaction can continue only until the capacity of the pore space is depleted relative to the hydrocarbon material.

The magnitude of this capacity is determined by the solubility of hydrocarbon in the pore water under specified pressure and temperature. The limiting value of the concentration of derivatives in the layer place express the maximum capacity of the reaction volume. The velocity of reaction considering this case, is limited by the introduction of (5.22) factor $\alpha = 1 - C / C_0$, where C - the current concentration of hydrocarbon in the pore water, C_0 - its maximum value.

Thus, the velocity of generation of hydrocarbon per unit area of the roof of OGMT is (actually generation process is implemented in OGMT thick h, but all products of this generation pass through the surface of contact generating thickness with the collector. Therefore, the amount of generation is convenient to normalize the surface area of the roof OGMT.)

$$V_G = \varepsilon h \Gamma (1 - C / C_0) e^{-\varepsilon t} \tag{5.23}$$

where;

h ; Thicknsse of OGMT

The magnitude of the current concentration of hydrocarbons in the flow of pore water, squeezed out from OGMT, by definition, is equal

$$C = \frac{x}{W} = \frac{h\Gamma}{W}\left(1 - e^{-\varepsilon t}\right) \tag{5.24}$$

where;

W; the volume of pore water, released as a result of compaction OGMT and/or passed through it over time t;

x; the mass of the produced hydrocarbon substances.

Substituting (5.23) and (5.24) (5.21), we have

$$G = \varepsilon h\Gamma \int_0^t \left[1 - \frac{h\Gamma\left(1 - e^{-\varepsilon t}\right)}{C_0 W}\right] e^{-\varepsilon t} dt \tag{5.25}$$

which after integration and simple transformations gives

$$G = h\Gamma\left(1 - e^{-\varepsilon t}\right)\left[1 - \frac{h\Gamma}{2WC_0}\left(1 - e^{-\varepsilon t}\right)\right] \tag{5.26}$$

According to (5.23) obtained formula remains valid until $C < C_0$. When $C = C_0$ then the hydrocarbon generation ceases. If the current hydrocarbon concentration exceeds a specified limit (i.e. hydrocarbon products of degradation are beginning to stand out in a separate phase, forming gas bubbles or droplets of oil) then the formula loses physical meaning.

Formula (5.26) is valid during the primary migration of hydrocarbons and loses its physical meaning in the transition to the process of secondary migration. From this it follows that the inequality $G > 0$ is always performed, $G = 0$ at $t = 0$ and when $t \to \infty$, G tends to $G_0 = h\Gamma$, which is quite realistic.

Since $W = w_0 + W_\phi$

Where;

W_ϕ; volume of failuation, which flow through unit area of OGMT by thickness h, and

w_0; Volume of pore fluid contained in a block of OGMT with a single base and height h, value $h\Gamma / 2C_0W$ is always less than 1. However, w_0 is the volume in which occurs in the reaction of degradation of DOM. It is equal to the volume of voids of OGMT, including the volume occupied by the DOM. This can be written as $W_0 = hn + h\Gamma / \rho$, which in turn suggests that the inequality

$$\frac{h\Gamma}{2C_0W} = \frac{h\Gamma}{2C_0\left(hn + h\Gamma / \rho + W_\phi\right)} < 1$$

It should be noted that the density of DOM does not exceed the $2/t/m^3$, and is often significantly less; $h/n > h\Gamma$ in most cases is of practical interest, and $2C_0$ at a depth of petroleum is always more than $1kg/m^3$.

As formulated in the task, the movement of pore fluid in the clay thickness in the inviolate natural environment under the principles of geofluid dynamics of slow flow is realized in thefiylation regime. It is absolutely clear that the alien pore water molecules of hydrocarbon - derivatives will be forced to fall on the axis of the pore channels with greater frequency than the water molecules. This is due to the fact that near the walls of the pore channels molecules of water are additionally bonded by surface forces of mineral skeleton, whereas in the center of the channel the resultant of these forces is equal to zero (Arie A.G., Slavkin B.C., 1995). Similar conclusion was obtained P.A. Diskej (Diskej P.A., 1975). Therefore, the concentration of hydrocarbons in failation flow is greater than C_0. However, the cramped conditions of the pore space in OGMT, as noted, do not allow such molecules to combine into a separate phase. Consequently, they are forced to migrate in a homogeneous unstable mixture with molecules of the pore water.

Based on the above, one can calculate the concentration of hydrocarbons in failation stream, if their concentration in the volume of pore space is equal to C. The ideology of this calculation is fairly obvious.

If the channel pore with radius R and the length l contains N molecules of a solvent such as water, we can write down

$$R^2 l = N \cdot \frac{4}{3}\pi r^3$$

where;

r; radius of molecules of the solvent, the total number of which is n_t at the fixed moment of time are located on the channel axis(i.e. there are always vacancies). Then $l = 2rn_t$, which after substituting into the initial equation gives

$$n_t = \frac{2}{3} N \frac{r^2}{R^2}$$

In other words, the number of molecules that fall at the same time on the axis of the pore channel, in R^2 / r^2 time smaller than two-thirds of their total number N in the channel. And because the frequency of contact of the hydrocarbon molecules with the axis of the pore channel is priorited compared with the water molecules, their concentration in this part of the pore space is equal to

$$C_\phi = C \frac{N}{n_t} = \frac{3}{2} C \frac{R^2}{R^2} \tag{5.27}$$

The resulting formula characterizes the concentration of hydrocarbon-substances in filation flow (eqution (5.27) holds for the reaction volumes commensurate with the volume occupied by the molecules of derivatives, which is characteristic of the pore space of clay OGMT).

A capillary pressure force directed against the forces of buoyancy occurs at the hydrocarbon - cluster contact with manifold overlapping tight layer and therefore;

$$K_{np} \leq \frac{n_n}{2} \left[\frac{gH}{\sigma} (\rho_0 - \rho) + \sqrt{\frac{n}{2K}} \right]^2 \tag{5.28}$$

where;

K_{np} and n_n; coefficients of permeability and porosity of the proposed tires respectively;

K and n; coefficients of permeability and porosity of the manifold respectively; σ; border tension between the hydrocarbon-phase and the reservoir water;

H; Depth of reservoir;

g; Acceleration of free fall;

ρ_0 ; Density of reservoir water;

ρ ; density of the hydrocarbon phase

From the known Arrhenius equation the coefficient of the reaction velocity of degradation is equal to

$$\varepsilon = Ae^{-E/RT} \tag{5.29}$$

where;

E ; The activation energy of degradation;

R ; Universal gas constant;

T; Absolute temperature of the system.

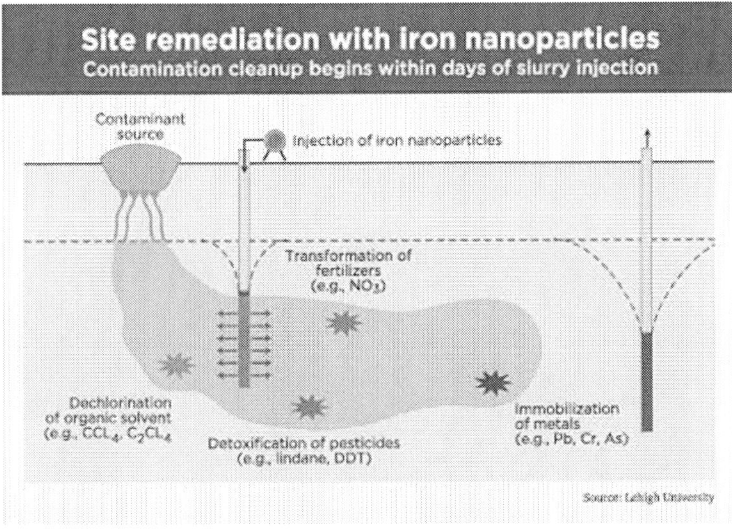

Figure 5.2. Cleaning the area with iron nanoparticles.

5.4. Properties of Aluminum

Aluminum powder reacts with water at 50° C with an allocation of hydrogen. It reacts vigorously in exothermic reactions with oxygen-containing liquids with halogenated organic compounds and other oxidants.

Aluminum is relatively nontoxic, relatively inexpensive, widely distributed in nature and in large quantities produced in industry by the electrolysis.

Aluminum in the form of nanopowder has lowered the ability to reaction at room temperature due to the presence of dense oxide-hydroxide shell, which is an electric double layer.

5.5. The Use of Aluminum Powder in Oil Production

Aluminium powder is used in the metallurgical industry in aluminothermy, as alloying additives for the manufacture of semi-finished products by pressing and sintering. Very strong components (gears, bushings, etc.) are received with this method. Powders are also used in chemistry for the preparation of compounds of aluminum as a catalyst (eg, production of ethylene and acetone).

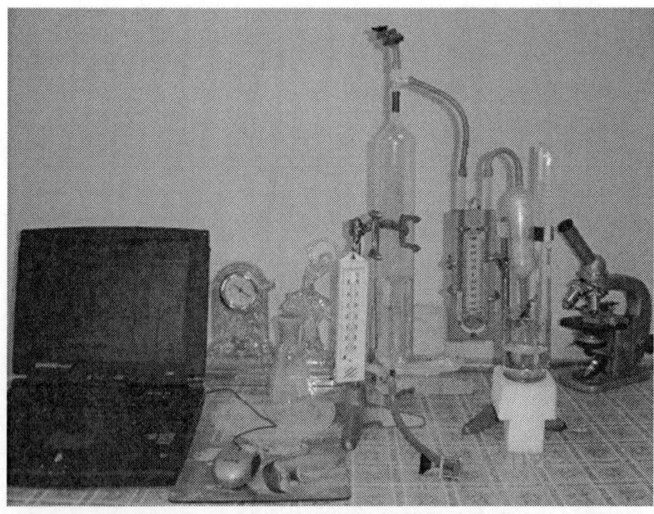

Figure 5.3. The picture of set, where measured the pressure.

Considering high reactivity of aluminum, especially in powder form, it is used in explosives and solid fuel for rockets, using its property to quickly ignite.

In the oil industry nanoparticles of aluminum were used for the separation of oil-water emulsion.

Figure 5.3 shows a snapshot of the device, where osmotic pressure was measured. The pressure difference Δp was measured with a liquid manometer - U -shaped tube. Manometer was connected to a closed receptacle, which was partially vacuumed to avoid the effects of atmospheric pressure Δp.

The use of aluminum powder in injection wells for water injection leads to a significant increase in reservoir pressure, creating the effect of hydrobreak and facilitates the efficient displacement of oil.

REFERENCES

Arie A.G., Slavkin B.C. About the mechanism of oil gas saturation of sand lenses. *Oil and Gas Geology*, 1995, № 2.

Bakajin Olgica, Duke Thomas A. J., Tegenfeldt Jonas, Chou Chia-Fu, Chan Shirley S., Austin Robert H., Cox Edward C. Separation of 100-Kilobase DNA Molecules in 10 Seconds, *Analytical Chemistry*; 73(24); 6053-6056 (2001).

Biomolecules with Microdevices", Electrophoresis, 21 (1), 81-90 (2000) Barrat J.-L., Chiaruttini F. *Mol. Phys*. 2003, 101, 1605–1610.

Bortov V.Yu., Garanin D.I., Georgievski V.Yu et.al. Comparative tests of reforming catalysts by "AKSENS". *Petrochemical and Refining*, 2003, vol. 2, p. 10-17.

Buyanov R.A., Krivoruchko O.P. Development of the theory of crystallization of sparingly soluble metal hydroxides and scientific basis of catalyst preparation of compounds of this class. *Kinetics and Catalysis*, 1976, v. 17, № 3, p. 765-775.

Cao G., Qiao Y., Zhou Q., Chen X. *Molecular Simulation* 2008, 88, 371–378.

Chernoivanov V.I., Mazalov Yu.A., Soloviev R.Yu etc. A method of making the composition. RF Patent on application № 2003119564 from 02.07. 2003.

Chemical Encyclopedia. Ed. I.L. Knunyants. Moscow: Soviet Encyclopedia, 1990, vol. 1, 2.

Chou C.-F., Austin R. H., Bakajin O., Tegenfeldt J. O., Castelino J. A., Chan S. S., Cox E. C., Craighed H., Darnton N., Duke T. A. J., Han J., Turner S. Sorting biomolecules with microdevices, <http://www.ncbi.nlm.nih.gov/pubmed/10634473##>Electrophoresis<http://www.ncbi.nlm.nih.gov/pubmed/10634473##>. 2000 , 21(1):81-90.

Churaev N.V., Ralston J., Sergeeva I.P., Sobolev V.D. Electrokinetic properties of methylated quartz capillaries. *Journal of Colloid and Interface Science*, 2002, V. 96, p. 265-278.

Cottin-Bizonne C., Barentin C., Charlaix E. etc. Dynamics of simple liquids at heterogeneous surfaces: Molecular-dynamics simulations and hydrodynamic description. *The European Physical Journal E*, 2004, V. 15, p. 427-438.

Cottin-Bizonne C., Cross B., Steinberger A., Charlaix E. Boundary Slip on Smooth Hydrophobic Surfaces: Intrinsic Effects and Possible Artifacts. *Physical Review Letters*, 2005, V. 94, p. 056102.

Culligan, Taewan Kim, Yu Qiao. Nanoscale Fluid Transport: Size and Rate Effects. *NANO LETTERS*, Vol. 8, No. 9, 2008, p. 2988-2992.

Darrigol O. Between hydrodynamics and elasticity theory: the first five births of the Navier-Stokes equations. *Archive for History of Exact Sciences*. 2002, v.56, p. 95-150.

Diskej P.A. Possible Primary Migration of Oil from Source Rock in Oil Phase // *Bull. AAPG*. – 1975. – Vol. 59, № 2.

Ermolenko, N.F., Efres M.D. Regulation of the porous structure of oxide adsorbents and catalysts. Moscow: *Nauka*, 1991.

Evstrapov A.A. The course of lectures "Nanotechnology in Environment and Medicine", 2011, 136 p.

Ficini J., Lumbroso-Bader N., Depeze J-K. Fundamentals of physical chemistry. – M., *Mir*, 1972.

Frenkel Ya.I. *UFN*, 1941, vol.25, no. 1, p. 1-18.

Friction laws at the nanoscale. *Nature*, 26.02.2009.

Gairik Sachdeva, Abhay Poal, Nelson Vadassery, Sayash Kumar, Srikar Chindada, Hanasaki, I.; Nakatani, A. *J. Chem. Phys*. 2006, 124, 144708.

Godymchuk A.Yu., Ilyin A.P., Astankova A.P. Oxidation of aluminum nanopowder in liquid water during heated. Proceedings of the Tomsk Polytechnic University, 2007, v. 310, № 1, p. 102-104.

Gusarov V.V., Popov I.Yu. Il Nuovo Cimento D, 1996, v.18, №7, pp. 799-805.

Hemanth Giri Rao. Will Water Flow Through Nanotubes? www.dstuns.iitm.ac.in/teaching-and-presentations/, 2010.

Holt J. K., Park H. G., Wang Y., Stadermann M., Artyukhin A. B., Grigoropoulos C. P., Noy A., Bakajin O. Fast Mass Transport Through Sub-2nm Carbon Nanotubes. *Science*, 2006, v. 312, p. 1034-1037.

Hongfei Ye, Hongwu Zhang, Zhongqiang Zhang, Yonggang Zheng. Size and temperature effects on the viscosity of water inside carbon nanotubes.

Nanoscale Research Letters 2011, p. 6-87, http://www.nanoscalereslett. com/content/6/1/87.

Huang C., Wikfeldt K.T., Tokushima T., Nordlund D., Harada Y., Bergmann U., Niebuhr M., Weiss T.M., Horikawa Y., Leetmaa M., Ljungberg M.P., Takahashi O., Lenz A., Ojamae L., Lyubartsev A.P., Shin S., Petterson L.G.M., Nilsson A. The inhogeneous structure of water at ambient conditions. *Proceeding of the National Academy of Sciences*; http:/www.pnas.org/content/early/2009/08/13/0904743106.abstract.

Hummer G., Rasaiah J. C., Noworyta J. P. Nature 2001, 414, 188–190.

Ilyin A.P., Gromov A.A., Yablunovsky G.V. About activity of aluminum powders. *Physics of Combustion and Explosion*, 2001, v. 37, № 4, p. 58-62.

Ilyin A.P., Godymchuk A.Yu., Tikhonov D.V. Threshold phenomena in the oxidation of aluminum nanopowders. Physics and chemistry of ultrafine (nano-) systems: Proc. VII All-Russian. Conf. Moscow: Moscow Engineering Physics Institute Printing, 2005, p. 178-179.

Jason K.H., Hyung G.P., Yinmin W., Stadermann M., Artyukhin A.B., Grigoropoulos C.P., Noy A., Bakajin O. Fast Mass Transport Through Sub-2-Nanometer Carbon Nanotubes. *Science*, 2006. 312, 1034. pp. 1034-1037.

John A. Thomas, Alan J. H. McGaughey, Ottoleo Kuter-Arnebeck. Pressure-driven water flow through carbon nanotubes: Insights from molecular dynamics simulation. *International Journal of Thermal Sciences*, 49, p. 281–289, 2010, journal homepage: www.elsevier.com/locate/ i j t s.

Joseph P., Tabeling P. *Phys. ReV. E*, 2005, 71, 035303.

Kalra A., Garde S., Hummer G. Osmotic water transport through carbon nanotube arrays. *Proeedings of the National Academy of Sciences of the USA*, 2003, v. 100, p. 10175-10180.

Kotsalis E.M., Walther J.H., Koumoutsakos P. Multiphase water flow inside carbon nanotubes. *International Journal of Multiphase Flow*, 30, 2004, p. 995–1010.

Kozlova E.G., Emelianov Yu.I., Krasiy B.V. etc. The new catalysts for reforming of gasoline with an octane rating of 96-98. *Catalysis in Industry*, 2003, № 6.

Korchagina Yu.I., Chetverikov O.P. Methods of assessing the generation of hydrocarbons produce oil. M.: Nedra, 1983.

Lauga E., Brenner M.P., Stone H.A. Microfluidics: the no-slip boundary condition in Handbook of Experimental Fluid Dynamics. New York: Springer, 2006.

Lauga E., Stone H.A. Effective slip in pressure-driven Stokes flow. *Journal of Fluid Mechanics*, 2003, V. 489, p. 55-77.

Li T.D., Gao J., Szoszkiewicz R., Landman U., Riedo E. *Phys. Rev. B.* 2007. 75. 115415.

Majumder M., Chopra N., Andrews R., Hinds B. *J. Nature* 2005, 438, 44–44.

Mazalov Yu.A. Patent RF № 2158396. *The method of burning metal fuels*. 2000.

Mirzadzhanzade A.Kh., Maharramov A.M., Yusifzade Kh.B., Shabanov A.L., Nagiyev F.B., Mammadzadeh R.B., Ramazanov M.A. Study the influence of nanoparticles of iron and aluminum in the process of increasing the intensity of gas release and pressure for use in oil production. News of Baku University. *Science Series*, № 1, 2005, p. 5-13.

Mirzadzhanzade A.Kh., Maharramov A.M., Nagiyev F.B., Ramazanov M.A. Nanotechnology applications in the oil industry. Proceedings of the II-nd Scientific Conference "Nanotechnology-production 2005", November 30-December 1, 2005 Fryazino., p. 47-52.

Mirzadzhanzade A.Kh., Maharramov A.M., Nagiyev F.B. On the development of nanotechnology in the oil industry. *"Azerbaijan's Oil Industry»*, № 10, 2005, p. 51-65.

Mirzadzhanzade A.Kh., Bakhtizin R.N., Nagiyev F.B., Mustafayev A.A. Nano hydrodynamic effects on the base use of micro embryonic technology. *"Oil and Gas Business"*, Volume 3, 2005, p. 311-315.

Mirzadzhanzade A.Kh., Shahbazov E.G., Shafiev Sh.Sh., Nagiyev F.B., Osmanov B.A., Mammadzadeh R.B. Nanotechnology in the oil and gas production: research, implementation and results. Book of abstracts. Khazarneftgazyatag - 2006. International Scientific Conference on October 25-26, 2006, p. 47.

Morten Bo Lindholm Mikkelsen, Simon Eskild Jarlgaard, Peder Skafte-Pedersen. Experimental Nanofluidics. Capillary filling of nanochannels. MIC – Department of Micro and Nanotechnology Technical University of Denmark, June 20th, 2005.

Nagiyev F.B. Nonlinear oscillations of gas bubbles dissolved in the liquid. Izv.AN Az.SSR, ser.f.-tech and math. *Science*, № 1, 1985. p. 136-140.

Nagiyev F.B., Khabeev N.S. Dynamics of soluble gas bubbles. Proceedings of the Academy of Sciences USSA, *Fluid and Gas Mechanics*, № 6, 1985, p. 52-59.

Nagiyev F.B., Mustafin R.Kh. Using of high technologges in oil production. Intensification of oil production with aid of nanohydrodynamic effects usage. Collection of thesis International workshop “Socio-economic aspects of the energy corridor linking the Caspian Region with E.U." Baku, Azerbaijan, April,11-12th, 2007, p. 28-37.

Natsuki T., Endo M., Tsuda H. *J. Appl. Phys*. 99 034311, 2006.

Natsuki T., Hayashi T., Endo M. *J. Appl. Phys*. 97 044307, 2005.

Navier C.L.M.H. Memoire sur les lois du mouvement des fluids. Mémoires Académie des Sciences de l'Institut de France. 1823, v.1, p. 389-440.

Neimark I.E. The main factors influencing the porous structure of hydroxide and oxide adsorbents. *Colloid Journal*, 1982, Volume 4, № 4, p. 780-783.

Nigmatulin R.I. Fundamentals of mechanics of heterogeneous media. M., *"Nauka"*, 1978, 336 p.

Nigmatulin R.I. Dynamics of multiphase media. Part I, M., *"Nauka"*, 1987, 464 p.

Popov I.Yu., Chivilikhin S.A., Gusarov V.V. Model of the structured liquid through the nanotube: http://rusnanotech09.rusnanoforum.ru/Public/LargeDocs/theses/rus/poster/04/Chivilikhin.pdf.

Ou J., Perot J.B., Rothstein J.P. Laminar drag reduction in microchannels using ultrahydrophobic surfaces. *Physics of Fluids*, 2004, V. 16, p. 4635-4643.

Ou J., Perot J.B. Drag Reduction and μ-PIV Measurements of the Flow Past Ultrahydrophobic Surfaces. *Physics of Fluids*, 2005, V. 17, p. 103606.

Press release on the website of the University of Wisconsin-Madison. Models present new view of nanoscale friction, 25.02.2009.

Proskurovskaya L.T. Physical and chemical properties of electroexplosive ultrafine aluminum powders: Dis. Ph.D., Tomsk, 1988, 155 p.

Ramazanova E.E., Shabanov A.L., Nagiyev F.B. Perspectives of nanotechnology method applications for intensification oil-gas production. Collection of thesis International workshop "Electricity Generation and emission trading in South Eastern Europe". Sofia, Bulgaria, 21 September, 2007.

Rothstein, J. P.; McKinley, G. H. *J. Non-Newtonian Fluid Mech*. 1999, 86, 61–88.

Semwogerere D., Morris J. F., Weeks E. R. *J. Fluid Mech*. 2007, 581, 437–451.

Skoulidas A. I., Ackerman D. M., Johnson J. K., Sholl D. S. *Phys. ReV. Lett*. 2002, 89, 185901.

Sokhan V. P., Nicholson D., Quirke N. J. *Chem. Phys*. 2002, 117, 8531–8539.

Sorokin V.S. Variational method in the theory of convection. *Applied Mathematics and Mechanics*. Volume XVII, 1953, p. 39-48.

Stepin, B.D., Tsvetkov A.A. Inorganic Chemistry. Moscow: Higher School, 1994, 608 p.

Suetin M.V., Vakhrushev A.V. Molecular dynamics simulation of adsorption and desorption of methane storage managed nanocapsules. "All-Russian Conference with international participation the Internet "From nanostructures, nanomaterials and nanotechnologies for nanotechnology ", *Izhevsk*, 08/04/2009, p. 112.

Sunyaev Z.I., Sunyaev R.Z., Safiyeva R.Z. Oil dispersions systems. -M.: *Chemistry*, 1990.

Tienchong Chang. Dominoes in Carbon Nanotubes. *Physical Review Letters*, 101, 175501, 24 October 2008.

Thomas John A. and McGaughey Alan J. H. Reassessing Fast Water Transport Through Carbon Nanotubes. *NANO LETTERS*, 2008, Vol. 8, No. 9, p. 2788-2793.

Thomas John A. and McGaughey Alan J. H. Water Flow in Carbon Nanotubes: Transition to Subcontinuum Transport. prl 102, *Physical Review Letters*, p. 184502-1-184502-4, 2009.

Uchic1 Michael D., Dimiduk1 Dennis M., Florando Jeffrey N, Nix William D. Sample Dimensions Influence Strength and Crystal Plasticity. *Science* 13 August 2004: vol. 305 №. 5686, pp. 986-989.

Wang C. Y., Ru C. Q., Mioduchowski A. *Phys. Rev. B* 72, 075414, 2005.

Wang Q., Varadan V. K. *Int. J. Solids Struc*. 43. 254, 2006.

Wei-xian Zhang. Nanoscale iron particles for environmental remediation: An overview. *Journal of Nanoparticle Research* № 5, pp. 323–332, 2003.

Whitby M. and N. Quirke. Fluid flow in carbon nanotubes and nanopipes. Chemistry Department, Imperial College, South Kensington, London SW7 2AZ, UK. *Nature nanotechnology*, www.nature.com/ naturenanotechnology, vol. 2, p. 87-94, February 2007.

Xi Chen, Guoxin Cao, Aijie Han, Venkata K. Punyamurtula, Ling Liu, Patricia J.

Yoon J., Ru C. Q., Mioduchowski A. *Compos. Sci. Technol*. 63, 1533, 2003.

Yoon J., Ru C. Q., Mioduchowski A. *J. Appl. Phys*. 93. 4801, 2003.

Yoon J., Ru C. Q., Mioduchowski A. *J. Composites* B 35, 87, 2004.

INDEX

A

accelerator, 43
acetone, 88
acid, 18
acoustics, vii
activation energy, 87
active compound, 74
additives, 8, 88
adhesion, 21
adsorption, xxii, xxiii, xxv, xxvi, 96
aggregation, 73
ammonia, 27
annealing, 18
architect, 6
aromatic hydrocarbons, 74, 75
Arrhenius equation, 87
atmosphere, xxi, 18
atmospheric pressure, 89
atomic force, 22
atoms, xi, xvi, xix, xx, xxii, 2, 3, 5, 6, 7, 11, 12, 17, 22, 25, 36, 41, 58, 73

B

bacteria, xxi
band gap, xvi, 18
bandwidth, xiii
base, xvi, xxii, xxiii, xxiv, 14, 74, 75, 85, 94
batteries, 8
benzene, 74
biomolecules, 91
births, 92
bloodstream, xii
bone, 43

C

capillary, xxiv, xxv, 8, 17, 18, 20, 81, 86
carbon, ix, xi, xii, xiii, xiv, xvi, xviii, xix, xx, xxi, xxii, xxiv, 1, 3, 4, 5, 6, 7, 8, 10, 12, 15, 18, 24, 34, 35, 36, 37, 38, 40, 41, 43, 44, 45, 46, 48, 49, 50, 53, 67, 68, 69, 72, 73, 74, 82, 92, 93, 96
carbon atoms, xix, xx, xxiv, 1, 3, 4, 5, 6, 7, 12, 15, 24, 41, 43, 44, 45, 73
carbon dioxide, xxi
carbon nanotubes (CNTs), xi, xii, xiii, xiv, xvi, xxi, xxii, 3, 4, 5, 8, 10, 11, 18, 34, 35, 37, 40, 43, 44, 48, 49, 50, 51, 52, 53, 54, 55, 56, 67, 68, 69, 92, 93, 96
catalyst, xiii, 88, 91
C-C, xix, xx, xxi
cell membranes, 22
characteristic path length, xiii
chemical, xiii, xv, xvi, xxi, xxii, 20, 25, 71, 73, 74
chemical properties, xxi, 75, 76, 77, 95
chemical stability, xvi

chemisorption, xxii
chirality, xvi, 10, 11, 12, 13, 14, 38, 45, 52
chlorophyll, 74
classes, 73, 74, 75
classical mechanics, 39
classification, 7, 74, 75
clay minerals, 71
clothing, xvii
clusters, 58, 59
coal, xxi
color, 74
combustion, xxii, 18
commodity, 78
compaction, 84
composites, 8
composition, xii, xvi, 60, 71, 73, 74, 91
compounds, xvi, 6, 72, 74, 88, 91
compression, 24, 44
computer, 47
computing, 48
conductivity, xv, xvi, 5, 18
conference, vii
configuration, 44, 45, 67, 68
conformity, 38, 64
constituents, 75
construction, xxi, 6
containers, xxii
Continuum Hypothesis, 20
convergence, 38
cracks, xvii, xviii, xxii
critical value, 31, 62
crystalline, 1, 7, 59, 73
crystallites, xv, 38
crystallization, 91
crystals, 1, 58
cycles, xxvi

D

DEA, vii
defects, xvii, xviii
deformation, xvii, xviii
degradation, 84, 85, 87
Denmark, 94
density fluctuations, xv
depth, xxvii, 71, 85
derivatives, 20, 83, 85, 86
desorption, xxiii, xxiv, xxv, xxvi, 96
destruction, xxvii
detection, 20
diameter of nanotubes, xiii, 5
differential equations, 20
diffusion, xv, 19
dispersion, 17
displacement, 71, 76, 77, 89
dissociation, xxii
distribution, 20, 25, 36, 37, 41, 62, 69
distribution function, 25
DNA, 91
drainage, 76
drawing, xvii
drugs, xi

E

Eastern Europe, 95
ecology, xxii
education, vii
educational process, ix
elastic deformation, xvi, 45
electric current, 19
electric field, xxiii, xxiv, xxv, xxvi, 19
electrical conductivity, xvi
electrodes, 5
electrolysis, 88
electromagnetic, xv
electromagnetic waves, 8
electron, xxvii, 1, 38
electrons, xv, xvi
electrophoresis, 19
emission, 18, 46, 95
energy, xix, xx, xxii, xxvi, 8, 20, 21, 22, 23, 43, 44, 45, 72, 79, 95
engineering, vii
environment, 85
environments, xxvi, 72
equality, 80
equilibrium, 24, 25, 41, 42, 43, 58, 59
etching, xiv
ethylene, 88

explosives, 89
external influences, 76
extraction, xxiv
extrusion, xxvi

F

fibers, 1
filters, xiv, xxi
filtration, xiv, 72
fluctuations, 21
fluid, ix, xi, xii, xiv, xv, xxvii, 17, 19, 21, 33, 35, 38, 39, 41, 47, 48, 49, 55, 58, 59, 60, 61, 62, 63, 64, 65, 67, 69, 70, 72, 81, 85
fluorescence, xii
force, xvii, xviii, xix, xx, xxi, xxvii, 17, 22, 23, 27, 47, 57, 61, 79, 82, 86
formation, xviii, xxii, 20, 36, 71, 78, 80, 82, 83
formula, xiv, 14, 39, 48, 49, 52, 61, 62, 63, 64, 68, 69, 73, 74, 80, 82, 83, 84, 86
free energy, 41
friction, 21, 47, 48, 61, 70, 72, 95
fuel cell, xxii
fullerene, 1, 6, 7, 45, 46

G

gas molecules, xiii, xiv, xxiii, xxv, 26, 59
geometry, 20, 45
global warming, xxi
graduate students, ix
grants, vii
graphene sheet, 3, 10
graphite, 1, 2, 3, 5, 6, 7, 8, 9, 12, 18
gravity, 23, 44, 57
grids, 1
growth, xxi, xxvii, 41
growth rate, 41

H

hair, 5
heat transfer, xv
height, 14, 33, 85
helium, 25
hemoglobin, 74
heterogeneity, 36, 37, 49, 58
high strength, xvi
higher education, ix
human, 5
hydrocarbons, 72, 73, 74, 77, 82, 83, 84, 85, 93
hydrogen, xxi, xxii, 8, 18, 24, 25, 36, 72, 73, 74, 88
hydrogen atoms, 36, 73
hydroxide, 88, 95
hypothesis, ix, 20, 21, 22, 48

I

ideal, xvi, 21, 26, 27
ideology, 85
image, 7, 21, 68
imagination, 45
incomplete combustion, xi
industrial emissions, xi
industry, 88, 89, 94
inequality, 84, 85
inertia, 22, 54
inhomogeneity, 78
institutions, ix
integration, 84
interface, 81
intermolecular interactions, 47
ions, 19
Iran, vii
iridium, xvi
iron, xvi, 74, 87, 94, 96
iteration, 64

K

kidneys, xii
kinetics, xxvii
Knudsen flow, xiv

L

laminar, 20, 21, 60, 62
laws, xv, 47, 48, 92
lead, 17, 35
leadership, xiii
legend, 52
light, xvii, 25, 74
liquid phase, xv, 42, 60, 80, 81
liquids, xi, xiv, xv, xxvii, 21, 34, 62, 73, 88, 92
lithium, 8
low temperatures, 25
lymph node, xii

M

magnetic properties, 8
magnitude, xii, xiv, xxv, xxvi, 17, 18, 23, 26, 35, 39, 41, 42, 77, 79, 83
majority, xxvi
manufacturing, xxvii
mass, ix, xv, xvi, xxi, xxiv, 14, 15, 19, 20, 21, 25, 28, 29, 39, 58, 59, 77, 79, 82, 83, 84
materials, vii, xvi, xxi, xxii, 47
mathematics, vii, viii
matrix, xiv
matter, 20, 58
measurement(s), xiv, xxvii, 21, 31
mechanical stress, xvii
media, vii, xxvii, 72, 95
membranes, xii, xiii, xiv, 35
Mendeleev, 26
metal hydroxides, 91
metal ions, xxi
metals, 5, 18
meter, xxii
microelectronics, xvi
microscope, 1
microscopy, xxvii
migration, xiv, 19, 84
models, xxvii, 20, 69
modifications, xiv, 20
modulus, xvi, 5
molar volume, 22
mole, xx, 26, 28
molecular dynamics, xxii, xxiii, 22, 25, 34, 43, 44, 47, 48, 59, 69, 93
molecular mass, 28, 74
molecular structure, 21
molecular weight, 27, 29
molecules, xi, xiii, xiv, xxi, xxii, xxiii, xxiv, xxv, xxvi, 6, 17, 19, 21, 22, 23, 24, 25, 26, 27, 34, 36, 37, 38, 39, 40, 41, 43, 47, 58, 59, 67, 68, 69, 70, 73, 79, 81, 82, 85, 86
momentum, 21

N

nanodevices, xxi
nanomaterials, 96
nanometer, xii, xv, xvi, xxi, 34
nanometer scale, xv
nanometer size channels, xii
nanometers, xii, xiii, 1, 3, 17
nanoparticles, xi, xii, 87, 89, 94
nanostructures, 3, 47, 96
nanosystems, 34, 35
nanotechnologies, xxvii, 96
nanotechnology, xi, xv, xxvii, 4, 7, 94, 95, 96
nanotube, ix, xii, xiii, xiv, xv, xvi, xviii, xix, xx, xxi, xxii, xxiv, xxv, 4, 5, 6, 7, 8, 9, 10, 11, 12, 13, 14, 15, 17, 18, 34, 36, 37, 38, 42, 43, 44, 45, 46, 48, 50, 51, 52, 55, 58, 59, 60, 61, 62, 63, 64, 65, 66, 67, 68, 69, 93, 95
National Academy of Sciences, viii, 93
natural gas, xv, xxii, xxiii
neglect, 72
nickel, 74
nitrogen, 34, 73, 74
nitrogen compounds, 74
nodes, xii, 1
nuclei, xx
nucleus, 63

O

octane, 93
oil, vii, xxi, xxvii, 71, 72, 73, 74, 75, 76, 77, 78, 81, 82, 83, 84, 89, 91, 93, 94, 95
oil production, 72, 94
opportunities, xxvii
organic compounds, 88
organs, xii
osmosis, 19
osmotic pressure, 89
oxidation, xxii, 93
oxygen, 36, 73, 74, 88

P

parallel, xxi, 1, 35
patents, viii
permeability, xxvii, 72, 87
petroleum, vi, 72, 73, 74, 80, 85
phonons, xv
photons, xvi
physical chemistry, 92
physical properties, xxvii, 73
physics, vii, viii, xiv, xv, xxvii
plants, 74
plastic deformation, xvii, xviii
polar, xii, 38, 74
pollutants, xxi
pollution, xxvii
polycarbonate, xiv
polymers, 8
porosity, 78, 87
porous materials, xiv, 71
porous media, 71
porphyrins, 73, 74
potassium, xxiii
power generation, 45
precipitation, xiii, xiv
preparation, 71, 78, 88, 91
President, viii
pressure gradient, 33, 49, 51, 52, 53, 54, 55, 56, 57, 72
principles, 20, 85
probe, xvi, 19, 20, 21
purity, xxvii

Q

quantum phenomena, xvi
quartz, 92

R

radius, xviii, 22, 36, 39, 40, 42, 54, 55, 60, 61, 62, 63, 64, 65, 66, 68, 69, 79, 81, 85, 86
RDP, 37
reactions, 45, 88
reactivity, 89
real-time fluorescence spectroscopy, xii
receptacle, 89
recovery, 18, 71
recovery processes, 77
relaxation, 25
remediation, 96
reprocessing, 74
repulsion, 23, 24, 38, 42, 67
researchers, xii, xiv, xv, 1, 36, 47
reserves, xxi, 75, 76, 77
resins, 75
resistance, xiv, 18
response, 19
restoration, 17
rings, xxiii, 45
rodents, xii
rods, 5
room temperature, 88
roots, 30
roughness, xiv, xvi, 21
Russia, vii, viii

S

safety, xxii
saturation, 91
schema, xix
sediment, 1

sedimentation, xxvii, 19
sediments, 1
selectivity, xiv
self-assembly, 5
semiconductor(s), xvi, 5, 11, 18
sensors, 8
shape, xvii, 7, 52, 69
shear, 17, 21, 38, 39, 41, 42, 62, 64, 65, 66, 69
signs, xxi
silica, 60
silicon, xii, xiv
simulation(s), ix, xv, 20, 34, 42, 43, 48, 52, 69, 93, 96
sintering, 88
skeleton, 85
smoothness, 67
sodium, 74
solid state, xvi, 73
solubility, 74, 83
solution, xxii, 64, 65
solvents, 74
Soviet Union, viii
specific surface, 78
spectroscopy, xii
sponge, xxii
stability, xxvii
stabilizers, 74
stable states, 43, 45
state, 26, 28, 31, 43, 58, 59
states, 31, 43, 44, 73
steel, xvi, xvii, xxi, 5
storage, xii, xxi, xxii, xxiii, xxvi, 96
stress, xvii, xviii, 38, 39, 40, 41, 42, 62, 64, 65, 66, 69
stretching, xxi
structure, ix, xi, xv, xxiii, xxvi, 1, 4, 7, 9, 11, 23, 35, 37, 38, 42, 43, 45, 60, 62, 64, 66, 67, 73, 92, 93, 95
substitution, 71, 80
substrate, xiv, 18
sulfur, 73, 74, 75
superconductivity, 8
surface area, xxi, xxii, 83
surface friction, 34
surface properties, xi
surface tension, 18, 78, 79
symmetry, 4, 5, 8
synthesis, xvi, 69

T

technologies, xvi, 76
technology, xxi, 77, 94
temperature, 25, 29, 30, 31, 34, 78, 79, 81, 83, 87, 92
tensile strength, xviii
tension, xvii, xix, xxvi, xxvii, 18, 24, 81, 87
test data, 42
thermodynamic parameters, xv
thermodynamics, xxvi
titanium, xvii
transformation(s), 8, 80, 84
transport, xii, xiii, xv, 20, 22, 37, 93
transport processes, 20
transportation, xxii
turbulence, xv

U

uniform, 63
universe, xxii
universities, vii
urban, vii

V

vacancies, 86
vanadium, 74
vapor, xiii, xiv, 73, 74
variables, 31
vector, 10, 11, 36
velocity, xiv, xxiv, xxv, 19, 20, 21, 25, 33, 34, 35, 37, 38, 39, 46, 48, 49, 52, 54, 55, 57, 60, 61, 62, 63, 68, 69, 82, 83, 87
viruses, xxi
viscosity, xv, xxvii, 20, 33, 38, 39, 40, 41, 42, 43, 48, 49, 50, 51, 60, 61, 72, 73, 92

W

water, xiii, xiv, xxi, xxii, 34, 35, 36, 37, 38, 39, 40, 41, 42, 43, 48, 49, 50, 51, 52, 54, 55, 58, 59, 65, 67, 68, 69, 71, 74, 75, 78, 81, 82, 83, 84, 85, 86, 87, 88, 89, 92, 93
water vapor, 34
wells, 71, 89
wires, 18

Y

yield, 45